THE 21ST CENTURY BRAIN

Other Books by Richard Restak, MD

Pre-Meditated Man

The Brain: The Last Frontier

The Self Seekers

The Brain

The Infant Mind

The Mind

The Brain Has a Mind of Its Own

Brainscapes: An Introduction to What Neuroscience Has Learned About the Structure, Function, and Abilities of the Brain

Receptors

The Modular Brain

Older and Wiser

The Longevity Strategy: How to Live to 100 Using the Brain-Body Connection (co-authored with David Mahoney)

Mysteries of the Mind

Mozart's Brain and the Fighter Pilot

Poe's Heart and the Mountain Climber

The New Brain

The Naked Brain

Think Smart: A Neuroscientist's Prescription for Improving Your Brain's Performance

The Big Questions: Mind

How to Prevent Dementia: Understanding and Managing Cognitive Decline

The Complete Guide to Memory: The Science of Strengthening Your Mind

THE 21ST CENTURY BRAIN

HOW OUR BRAINS ARE CHANGING IN RESPONSE TO THE CHALLENGES OF SOCIAL NETWORKS, AI, CLIMATE CHANGE, AND STRESS

RICHARD RESTAK, MD

Skyhorse Publishing

Copyright © 2025 by Richard Restak, MD

All rights reserved. No part of this book may be reproduced in any manner without the express written consent of the publisher, except in the case of brief excerpts in critical reviews or articles. All inquiries should be addressed to Skyhorse Publishing, 307 West 36th Street, 11th Floor, New York, NY 10018.

Skyhorse Publishing books may be purchased in bulk at special discounts for sales promotion, corporate gifts, fund-raising, or educational purposes. Special editions can also be created to specifications. For details, contact the Special Sales Department, Skyhorse Publishing, 307 West 36th Street, 11th Floor, New York, NY 10018 or info@skyhorsepublishing.com.

Skyhorse® and Skyhorse Publishing® are registered trademarks of Skyhorse Publishing, Inc.®, a Delaware corporation.

Visit our website at www.skyhorsepublishing.com.
Please follow our publisher Tony Lyons on Instagram @tonylyonsisuncertain

10 9 8 7 6 5 4 3 2 1

Library of Congress Cataloging-in-Publication Data is available on file.

Cover design by Brian Peterson
Cover image by John Wylie and ChatGPT

Print ISBN: 978-1-5107-8415-4
Ebook ISBN: 978-1-5107-8416-1

Printed in the United States of America

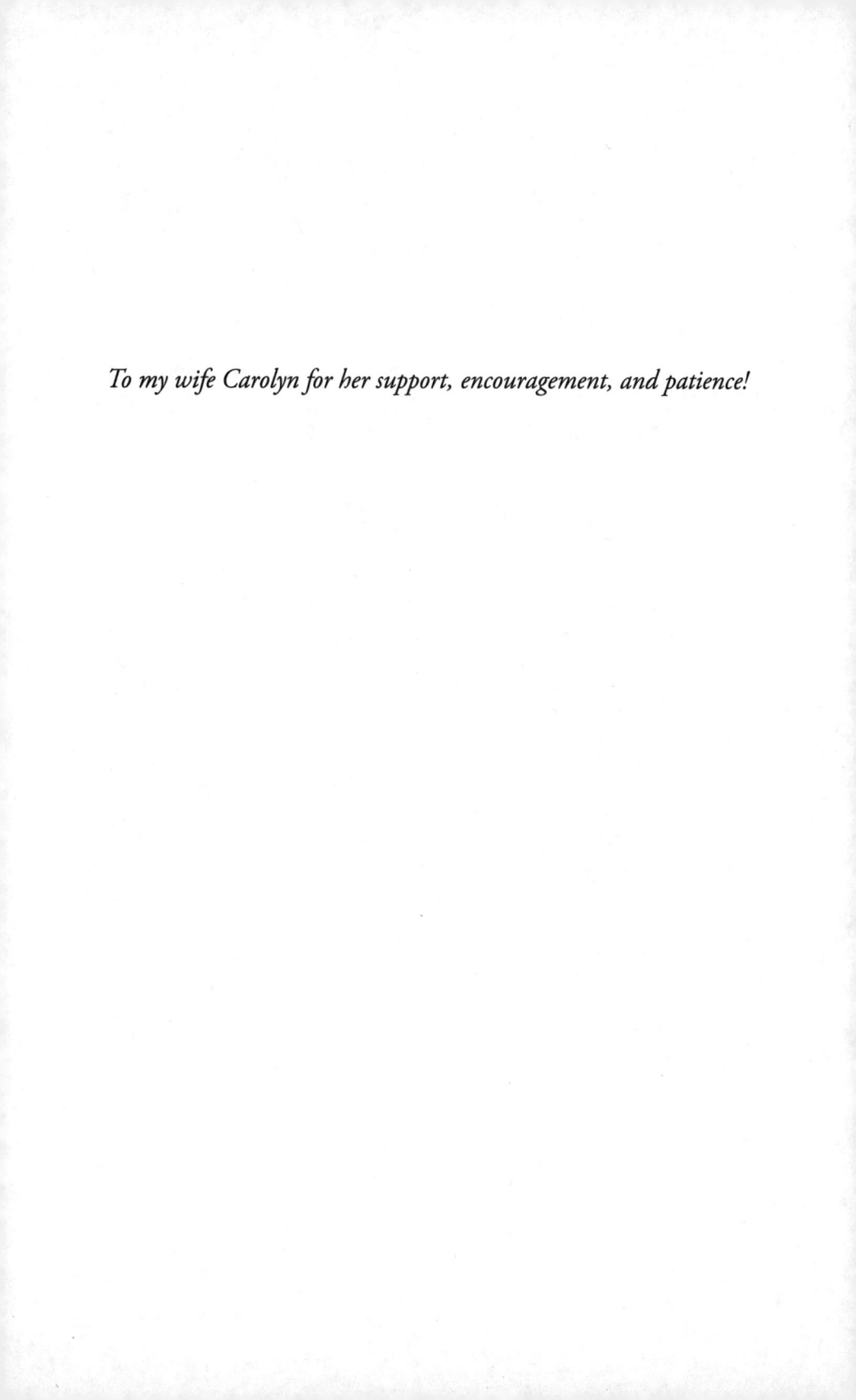

To my wife Carolyn for her support, encouragement, and patience!

Contents

I. Introduction — 1

II. Twenty-First-Century Brain Disorders — 7
1. COVID-19, the virus that just won't say goodbye — 7
2. New, more dangerous infectious agents — 9
3. Dr. Fauci catches the bug — 10
4. Binge drinking in plastic cups — 12
5. Only a fraction of the size of a human hair — 18
6. "The Hot New Luxury Good for the Rich: Air" — 20

III. Global Warming — 23
1. "Travel prepared to survive" — 23
2. Heat and the frontal lobes — 25
3. Lifeboat mentality — 29
4. Heat-fueled aggression — 32
5. The Keeling Curve — 33
6. Everything all at once — 37
7. A triumph — 45
8. Unanticipated consequences — 47
9. A David Lynch–inspired landscape — 51
10. "Only a rational business decision" — 53
11. The unwilling or unable — 55

IV. The Twenty-First-Century Brain — 59
1. Probabilistic conclusions — 59

 2. A huge bowl of spaghetti   61
 3. How to think about eighty-six billion neurons   62
 4. The connectomic brain   65
 5. Eleanor Maguire and the cabbies   67
 6. The most influential brain concept of the twenty-first century   71

V. The Internet   75
 1. Living in a bubble   75
 2. "Please don't take my internet away"   77
 3. "Not some isolated incident"   78
 4. "Oops, that sounds bad"   79
 5. A social media trap   81
 6. Is the internet addicting?   83
 7. Internet-stoked violence   87
 8. The internet's darkest corners   90

VI. Artificial Intelligence (AI)   95
 1. Elaine Herzberg's last shopping trip   95
 2. AI versus the brain   98
 3. What's the world record for crossing the English Channel entirely on foot?   100
 4. Rise of the hypnocracy   107
 5. Welcome to the "reality" of AI   111
 6. More existential concerns   113
 7. "Generative AI took over my life"   118
 8. The dark side of AI   122
 9. Where does paranoia begin?   124
 10. AI in the twenty-first century   125
 11. Eliza returns   128

VII. Misinformation and Disinformation   133
 1. The Doomsday Clock   133
 2. Prestige, fame, and potential fortune   137
 3. Dealing with doctors in unaccustomed ways   140
 4. Dr. Victorson's dilemma   144
 5. Welcome to scam world   146

VIII. Memory   149
1. "The past is a foreign country. They do things differently there"—L. P. Hartley   149
2. The war against the past   149
3. "Not for the faint of heart"   153
4. Minnie Spotted-Wolf defers to Elvis Presley   155

IX. Surveillance and Its Effects   161
1. Why is that library camera directed toward me?   161
2. "The sentiment of invisible omnipresence"   166
3. "Panopticons on wheels"   169
4. Smart glasses   176
5. "Maybe my neighbor really is a spy"   177
6. Surveillance as an exercise of intimidation   180

X. Anxiety   185
1. An inescapable aspect of our lives   185
2. The chronic fallback emotional state   187
3. Brain and anxiety   188
4. "Helpless against something I can't put into words"   192
5. "If it bleeds, it leads"   193
6. A child hit by a car   195
7. Traumatizing a population   197
8. Can nuclear war be avoided? For how long?   198
9. LA fires and Hiroshima   199

XI. New Ways of Thinking   203
1. Searching for hidden links among seemingly unrelated topics   203
2. The need for a "mental upgrade"   206
3. Occam was wrong   211
4. Volatility, uncertainty, complexity, ambiguity   214
5. A sensible solution   217

Acknowledgments   223

Principal Sources Consulted   225

CHAPTER ONE

Introduction

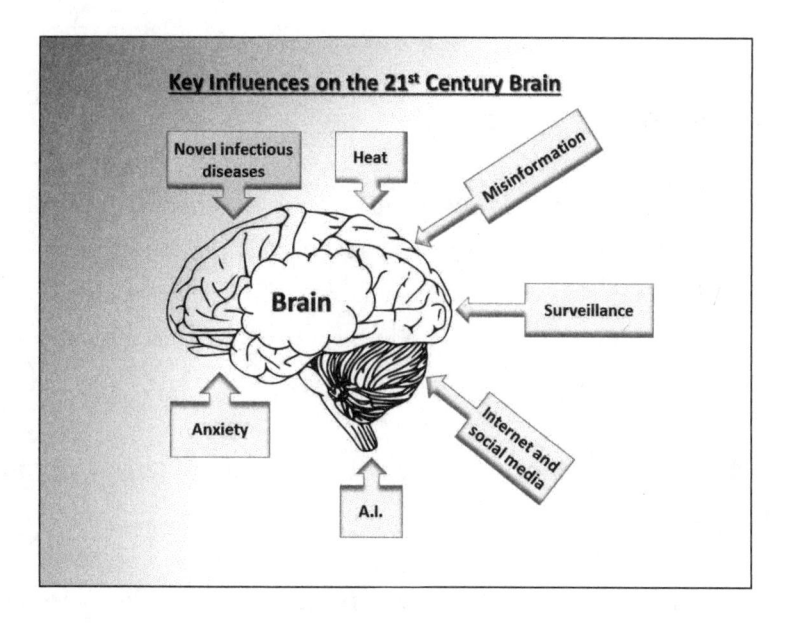

Key Influences on the 21st Century Brain

- Novel infectious diseases
- Heat
- Misinformation
- Surveillance
- Internet and social media
- A.I.
- Anxiety

→ Brain

As part of presenting the essential message of a book to the reader, an author is often asked to imagine taking a short trip on an elevator and during that two- or three-minute ride providing a hypothetical passenger with a summary of what the book is all about. Here is my "elevator pitch" for *The 21st Century Brain*:

All of the challenges discussed in this book have the potential, if unheeded, of creating destructive effects on brain function. As a consequence, the organ becomes less able to address the problems created by these challenges, as well as evoking the necessary thought functions required to resolve them. Air pollution and global warming, for instance, provoke declines in concentration, attention, working memory (keeping multiple thoughts simultaneously in mind), abstraction, and emotional expression and control. Further, as its dysfunction accelerates, the brain's ability to resolve these challenges diminishes in step with that acceleration. While there is still time to greatly improve the situation, the required solutions are likely to come from unexpected and even surprising sources, starting with the ways we conceptualize the problems.

Think of it as a two-way interaction. The challenges can harm the brain; if that happens the damaged brain is thereby limited in its power of perceiving resulting problems along with summoning the time and urgency required to come up with the solutions.

The 21st Century Brain explores challenges that are largely unique to this century.

Included are:

New and more deadly diseases—defined as either diseases restricted to the twenty-first century, or the more powerful expression of previously existing diseases. Included here are: COVID, which over a few months or weeks can produce the equivalent of twenty years of brain aging and, after its initial flare-up, often reverts to an enervating form known as long COVID; future pandemics, which researchers assert are more a matter of "when" rather than "if"; Lyme disease marked by an increase in infectivity along with a more panoramic array of symptoms affecting thinking, emotional control, concentration, and attention; and finally, one of the more recently noted brain afflictions, plastics, which according to recent research are a major contributing factor to dementia.

Global warming—has already culminated in deaths resulting from record-breaking temperatures along with impairments in the logical thinking and mental processing required to tackle these climate extremes. Thus we are now facing a unique situation: The exploratory, knowledge-seeking tool in question, *the brain*—the very organ employed for coming up with solutions to twenty-first-century challenges—may ultimately prove ineffective unless changes in its responses can be induced. More about that throughout the book.

The internet—despite the indisputable benefits conferred by the internet, it has from its inception engendered such counterproductive and insidiously harmful effects as misinformation, addictive social networks, social isolation, and violence.

Artificial intelligence—AI is especially challenging as a result of its ability to create synthetically generated text or images that, currently, humans have trouble distinguishing from the real thing. Added to this is AI's facility at disseminating misinformation; its tendency to breed dependency; and the formation of bizarre relationships such as those involving humans attracted toward "an online romance" with avatars.

Together the internet and AI have greatly contributed to the house-of-mirrors world we currently inhabit. Misinformation and disinformation didn't originate in the twenty-first century but can be traced back hundreds of years. Nonetheless, the extent and intensity of deceptive techniques leading to misinformation and disinformation only came to their full fruition in the twenty-first century. As a consequence, the brain faces a dilemma: Its abilities to make decisions and exercise judgments require access to reliable information. Lacking this, the brain must resort to guesses and suspicions, often leading to cultlike behavior culminating in outright paranoia. Currently the human brain is forced to operate tentatively, a modus operandi that leads to a simmering sense of uncertainty and anxiety.

Surveillance—Surveillance techniques have become such a powerful influence on our lives that once we leave our house or apartment, we can never be completely certain that we are not appearing on a surveillance camera somewhere. If you live in New York City, reportedly over fifteen thousand police surveillance cameras transfer images to facial recognition software. (Most experts I've spoken to believe the number is much higher.) Currently surveillance cameras have become so much a part of everyday life that the concept of public space has gradually expanded to include anywhere outside the strict confines of one's home—unless a person chooses to create a "wired home" in which just about every device is connected to the internet.

Anxiety—The culmination of the previously mentioned challenges has created the most anxious generation in history, far greater than the so called Age of Anxiety (1950s). Currently our entire culture is experiencing a sense of fragility, threat, and peril starting with the September 11, 2001, terrorist attack on New York City and the Pentagon and extending to the latest wildfire breakout. Increasing numbers of people are expressing fears that everything may soon completely spin out of control with the explosion of one or more atomic bombs.

In the *Bulletin of the Atomic Scientists*, Richard P. Turco and Owen Brian Toon, both distinguished scientists in atmospheric and oceanic sciences, wrote "Manmade climate change represents the slow withering of humankind. Nuclear war represents a swift annihilation, a vision of Armageddon." In addition to nuclear weapons, each of the other existential threats—future pandemics, climate change, habitat and diversity loss (nature crisis), and unregulated AI and the future consequences that may ensue—interact with each other. None can be considered in isolation.

A few words about the structure of the book. The challenges faced by the twenty-first-century brain don't come nicely packaged or easily distinguished. Nor are the consequences always predictable. Most frustrating of all, these challenges are all occurring simultaneously.

For instance, as I'm writing this, the midwestern United States is experiencing widespread tornadoes and flooding beyond anything recorded in decades; active wildfires continue to burn in eight states, including North Carolina, where the largest fire at Table Rock is only 30 percent contained; measles is expanding at near-epidemic proportions with increasing numbers of children affected in Texas, New Mexico, and Oklahoma, with the largest outbreak in West Texas; and widespread panic is reflected in the stock market, in response to the imposition of tariffs earlier this morning, which have already led to more than a 4 percent loss of the S&W—one could go on.

Constructing a narrative of these forces and how they affect the formation and function of the twenty-first-century brain demands an approach capturing how things are actually happening rather than forcing them into a linear narrative. Events are not occurring in a linear manner. To pretend otherwise is to lessen the chances of understanding and coming up with solutions to challenges to which the brain has never before been exposed. Throughout this book and especially in the final chapter, I'll provide suggestions for ways of conceptualizing and responding to these uniquely twenty-first-century challenges.

CHAPTER TWO

Twenty-First-Century Brain Disorders

(1) COVID-19, the virus that just won't say goodbye

Twenty-first-century disease has taken on a new dimension by attacking the brain in novel ways. Traditionally infectious diseases like the common cold and the flu follow a similar pattern: infection, illness, recovery. But with COVID-19 everything changed. In many cases, its symptoms may never go away during the remainder of the infected person's life. These include what patients and doctors alike refer to as long COVID or "brain fog": the vague, difficult-to-describe loss of mental clarity, logical and consecutive thinking, concentration, memory, mental imagery, the ability to think abstractly such as forming analogies and using metaphors, multitasking, synthesizing information, making deadlines, and tethering the mind to a specific train of thought.

While many of the worst residual symptoms occur among people who were afflicted with the most serious illnesses, residual symptoms of equal severity also occur with illnesses mild enough to require little or no treatment other than a few days at home self-medicating with fluids and over-the-counter medications. Researchers now believe long-term consequences of "long COVID" are the result of a viral injury to the brain wreaking global dysfunction across one or more areas of cognition. The resulting "brain fog" has wrecked careers and marriages. Difficulty with concentration and focus often decreases professional performance to unacceptable levels; unrelenting complaints of fatigue are often accompanied by inability or unwillingness to join the spouse in accustomed leisure activities, thus delivering a death blow to many marriages. "She never wants to do anything or go anywhere. All she wants to do when she gets home from work is eat dinner too quickly and go to bed" is a typical complaint that I've heard explaining why one spouse has decided to call it quits in response to the partner's long COVID.

Especially at risk for COVID damage is the gray matter of the brain, which is responsible for information transfer. Heightened inflammation affects primarily the central nervous system, similar to what is encountered with brain trauma. While some treatments used for brain injuries (speech, cognitive, and occupational therapies) have helped, they rarely cure the symptoms of "brain fog."

In terms of severity, the brain deficits found in COVID-19 are equivalent to twenty years of brain aging. Things are even more perilous in those who experienced early dementia prior to COVID-19. According to the National Institute of Neurological Disorders and Stroke (NIND&S), in those already afflicted with dementia, "the virus rapidly accelerated structural and functional brain deterioration." Nor should this come as a surprise. We already know that dementia is regularly associated with brain inflammation; some researchers are even convinced that underlying inflammation is the *primary* cause of Alzheimer's. To them the telltale amyloid plaques and tangles typical of the disease serve only as collateral damage developing after the initial inflammatory insult. *COVID-19 inflammation* may be the initial salvo fired at a person who shows no prior clinical signs or laboratory evidence of dementia.

As Ziyad Al-Aly, a brain researcher at Veterans Affairs St. Louis Health Care, thinks of it, "It will continue to infect the population. So if this is indeed a virus that damages the brain in the long term or permanently, we need to figure out what can be done to stop it." Until that goal is achieved, the brain is potentially in peril. Even more worrisome is this question: Is COVID only the first of a series of infections yet to come that are serious enough to compromise both the structure and the function of the twenty-first-century brain?

(2) New, more dangerous infectious agents

With increasing global temperatures, mosquitoes and other disease-causing organisms trek north in quest of their ideal temperatures. As a result, illnesses such as West Nile virus (WNV), Dengue fever, and Lyme disease have either appeared where they hadn't been encountered before or are common at sites where they formerly posed less of a threat. Added to this, some of the pathogens adopt, change, or evolve in response to temperature elevations. To take one example among many, under conditions of extreme heat, the *Culex mosquitoes,* the vectors (transmitters of pathogens) for West Nile virus, multiply more frequently and live longer. The end result is an expansion in humans of West Nile virus vulnerability.

While mild and even asymptomatic cases of WNV can occur, disabling and even fatal brain diseases are frequently encountered. But whether the illnesses are mild or deadly, the incidence of WNV is steadily increasing. According to the European Center for Disease Prevention & Control (ECDC), more than thirty

European regions have experienced locally acquired infections for the first time ever. The ECDC estimates that with further climate-induced expansion of the *Culex mosquitoes*, an additional 244 million European citizens will be at risk for contracting West Nile virus.

Higher temperatures do not just affect pathogen-mediated human illnesses. Genetic mutations may endow an infective organism with increasing tolerance for higher temperatures, thus enabling the organism to retain or even strengthen its potency and the variety of its manifestations. The increase in expansion and potency of Lyme disease spiked between 2013 and 2022 in comparison to earlier decades.

I can personally attest to the enhanced destructiveness of current Lyme disease compared to decades ago when I was in medical school. A very close friend contracted Lyme in 2022 while working over a two-year period off Cape Cod. Starting with an onset of sudden symptoms (an elevated temperature, headaches, fatigue, and general malaise), the illness took on a more indolent course. Over the following two years the headaches improved, but there has not been much improvement in general overall fatigue or the loss of energy needed for carrying out physically or mentally demanding activities. At first she was forced to discontinue marathon running and eventually any running at all. This progression of the disease starkly contrasts with the typical case of Lyme disease contracted a decade or so ago, which typically responded to a ten-day course of antibiotics.

(3) Dr. Fauci catches the bug

As a result of temperature-propagated migrations of creatures ranging in size from insects to humans, the incidence of infections and death have recently climbed, often secondary to rarely encountered diseases such as mosquito-borne encephalitis. In the summer of 2024, the extremely rare eastern equine encephalitis (EEE) began affecting New Englanders. Ordinarily the disease is rare with only eleven cases of EEE typically reported in the United States annually. But the disease can be a serious one since about 30 percent of people infected with EEE die, with many of the survivors forced to live with ongoing neurological problems.

The West Nile virus, also transmitted through the bite of an infected mosquito, is currently the leading mosquito-borne disease in the continental United States. A chain of infection passes from infected birds to mosquitoes that then feed on people, who in many cases are not aware that they have been bitten. Human infections are the end of the line for the virus, since humans do not develop high enough levels of the virus in the blood to pass it on to others.

These virus outbreaks are the result of rising global temperatures in more than two thirds of the United States that increased the number of "mosquito days," defined as days with an average humidity of at least 32 percent and temperatures between 50 and 95 degrees Fahrenheit. Since the Northeast has warmed at a particularly fast rate, it has experienced the largest increase in mosquito days. In Massachusetts the average increase has been fourteen more mosquito days compared with the period from 1980 to 2009. And if global warming continues to increase, the resultant longer mosquito season could boost the risk of outbreaks of diseases carried by bloodsucking insects like those responsible for dengue and malaria. Soon diseases we only hear mentioned in movies set in tropical climates may become incorporated into our everyday vocabulary.

In a cruel irony, Anthony Fauci, the former director of the National Institute of Allergy & Infectious Diseases, and the widely recognized face of the US government's response to the COVID pandemic, was bitten by a West Nile–infected mosquito in the summer of 2024 while he was gardening in his yard. Within days, Fauci developed fever, chills, and fatigue and was hospitalized for about a week. The initial bite occurred in Washington, DC—about three blocks from where I'm living. I mention this last point for a reason. Upon learning of Fauci's illness, I felt incredibly vulnerable. "Perhaps it wouldn't be such a great idea to continue regularly eating breakfast and dinner on the patio," I suggested to my wife.

Further, I became hyperalert for any insect that looked even remotely like a mosquito. Despite reading more about the illness in professional journals and available news sources, I continued to be uneasy. But no matter how much I read, there was no alternative to accepting my inability to definitively eliminate the possibility that I and others around me might become the next encephalitis victim. My brain—the twenty-first-century model—must learn to work with such underlying uncertainties.

(4) Binge drinking in plastic cups

According to findings from the National Centers for Health Statistic's National Health Interview Survey, COVID led to an increase in alcohol overuse along with other substance abuses. Forced to isolate in 2020, many people experienced stress, a sense of isolation, and uncertainty. They were at home during hours they would ordinarily be at work; in the absence of childcare some parents were forced to spend more time with their children than was good for either parent or child. In response to this stress, drinking was a socially acceptable response. Unfortunately some parents' exposure to alcohol progressed from end-of-the-day rituals to frequent "pick-me-ups" irregularly spaced throughout the afternoon and into the evening. Initially slight hangovers of various degrees inevitably led to various and varying signs and symptoms of withdrawal, marked by early-morning irritability, impatience, temper flare-ups, etc.

In response to these stresses, binge drinking and overall heavy drinking increased from about 5.1 percent of Americans in 2018 to 6.29 percent in 2022. Alcohol-related deaths also surged.

In 2020 alone, according to one study, a 25 percent increase occurred in heavy drinking in both men and women, across all ages, races, and ethnic groups.

Of the many people who reported heavy drinking (defined for men as at least five drinks per day or at least fifteen drinks per week; for women defined as four drinks a day or at least eight per week) two groups stand out: Adults in their forties reported the highest level; with adults in the fifty–sixty range following close behind. Among women of all ages the heavy drinking rate of 6.4 percent exceeded the 6.2 percent among men.

Overall, based on the COVID experience, we can anticipate national increases in alcohol and drug abuse in response to future pandemics.

Odds are that however much you were drinking during the pandemic, you were drinking at least occasionally from a plastic bottle. But that may not be a healthy habit.

Early animal studies pointed to impaired learning and memory associated with microplastics (MPs) ingestion. Among humans the evidence was initially suggestive rather than conclusive, but still concerning.

When researchers turned toward humans and examined brain samples from patients who had died from Alzheimer's disease, their brains contained up to ten times more plastic by weight than samples collected from people who were never diagnosed with Alzheimer's or any other degenerative disease of the brain. Even the background level of microplastics in the normal brain in 2024 was 50 percent higher than brain samples taken in 2016.

Humans are principally exposed to microplastics—fragments smaller than five millimeters in diameter—thanks to widespread plastic pollution in air, water, and food. The world is currently producing nearly half a billion tons of plastic per year; one garbage truck worth of plastic is dumped into the oceans every minute, resulting in a serious disruption of the marine ecosystem. For instance, a broad range of ocean and freshwater species become entangled in microplastics (MPs) and are subsequently ingested by fish; eventually these aquatic combinations of fish and MPs pass further along the food chain and end up in our stomachs. If this wasn't worrying enough, the microplastic particles are often contaminated by toxic chemicals and can serve as delivery systems of the toxins to the brain.

In an already classic population study, researchers at the Department of Pathology at the University of São Paulo in Brazil carefully examined the olfactory bulb tissues from the noses of fifteen deceased local residents ranging in age from thirty-five to a hundred years (one centenarian). Microplastics could be seen in eight of the fifteen brains—including the centenarian.

These findings included sixteen synthetic polymer particles commonly used in a broad range of products, including food packaging, textiles, kitchen utensils, medical devices, and adhesives. These findings are especially ominous when put into the context of the most frequent early symptom and sign of Alzheimer's disease: anosmia (loss of smell).

Size-wise these microplastic particles ranged in length from 5½ to 26 microns (a micron is one millionth of a meter). Width varied from 3 to 25 microns. The mean fiber length and width was 21 and 4 microns. To put these numbers in perspective, the diameter of a human hair is much larger, averaging about 70 microns.

Since the microplastics are smaller than a human hair, they have no difficulty after inhalation in entering the brain along the olfactory pathways. In earlier studies, particulate plastic inhalation was associated not only with dementia, but with other neurodegenerative diseases like Parkinson's disease. But the most worrisome implication of the São Paulo findings is the high likelihood that we may be inhaling microplatics while indoors. Indeed, this may be much more common than anyone previously imagined prior to the São Paulo study.

As Thais Mauad, associate physician, researcher, and senior investigator, explained it: "Breathing within indoor environments could be a major source of plastic pollution." What's more, this has to be put into the context of an absence of any current regulation. Federal, state, and local regulations for microplastics are "virtually non-existent," according to the Interstate Technology & Regulatory Council.

In November 2024, an Intergovernmental Negotiation Committee met in Busan, South Korea, in search of a legally binding global treaty on plastic production. While a hundred nations voted for a decrease in production of plastic, a handful of oil-producing nations put the kibosh on any hopes of a legally binding global treaty. The final agreement? Postpone negotiations to a later date. Now put this decision into the context of predictions that hold that if the current production of plastics isn't halted or drastically reduced, global plastic production can be predicted to triple by 2050.

As Panama's delegation head, Juan Carlos Monterrey Gomez, angrily retorted, "Every day of delay is a day against humanity. Postponing negotiations does not postpone the crisis."

Given such a disparity in the estimate of what will be required and who will have to provide it, one could be forgiven for thinking in terms of two different groups providing cost estimates for different problems. And in a sense, we are speaking of two different groups defined by the financial reserves each has available to combat both the plastic problem and climate change in general.

Simply put, some parts of the world (Think: the United States) aren't *yet* facing the same degree of threat as others.

We've witnessed this moral and ethical myopia before. Our political leaders and health czars were initially convinced that COVID only threatened "other parts of the world" until, as a nation, we ourselves experienced the tsunami of sickness, death, and extended negative outcomes (long COVID).

Despite recommendations to the contrary by experts on the chemistry of plastics, I'd wager you a modest bet that you are doing one or more of these medically inadvisable practices:

1. Refilling simple-use plastic water bottles despite evidence that the inside of the bottle sheds micro- and nanoplastics into the refilled water.
2. Bringing home a take-out meal in a plastic container and either eating it in the container or reusing that container to store leftovers. Such practices ignore findings that plastics contain about 16,000 chemicals, including styrene and about 4,000 others considered "highly hazardous." Black plastic is especially treacherous since it contains high levels of flame retardants in addition to the tiny nanoplastic and microplastic fragments found in all plastic containers.
3. Heating food in plastic containers labeled "micro friendly" or other designations suggesting potential reuse. This practice flies in the face of the most hard-and-fast rule about plastic: Never microwave plastic or put it into a dishwasher, especially when running the washer on a hot cycle.

While eliminating plastics would seem to be a sensible goal for everyone affected by the problem, the oil- and gas-producing countries (Saudi Arabia, Russia, and, yes, the United States) remain staunchly opposed to any regulation. Why? Because these countries produce the fossil fuels used to make plastics. Unless you are a chemist, you may not have been aware of this link between fossil fuels and plastics.

Despite the disturbing issues raised by unchecked levels of plastic production, the Food & Drug Administration (FDA) claims on its website that current research and scientific evidence does not demonstrate that the levels of MPs or nanoplastics detected in food preparation pose a risk to human health. Overall the FDA's reassurance rings hollow in my mind. Best to reduce your exposure to MPs by avoiding plastics in food preparation, especially microwaving, and replacing bottled water with tap water from a reliable source.

Not surprisingly, the environmental toll of plastics has grown in tandem with efforts by the plastics industry to neutralize critics. The industry group spokesman speaking for the plastic manufacturers NAPCOR (the National Association for PET Container Resources; PET stands for polyethylene terephthalate) claims that the majority of single-use plastic containers are recycled (untrue, with less than 30 percent representing the true figure, which translates into 70 percent either incinerated or carried by ocean currents to some of the beaches found in the poorest countries in the world). The spokesman also claims that plastic pollution is coming under control (also untrue and totally dismissive of the California attorney general's assessment when filing a lawsuit against Exxon Mobil for what was characterized as "a decades long campaign of deception").

NAPCOR's most recent and perniciously subtle approach involves the hiring of dozens of social media influencers as part of a Positively PET campaign aimed at convincing the general public that there is nothing to fear from plastics.

In the campaign, the real determining issue is conveniently omitted: Without fossil fuels, plastic production plummets and money is lost for oil and gas investors—the ultimate no-no in a country where, it seems, investor losses must be avoided at all costs, even if those costs involve poisoning the planet.

What's needed in the twenty-first century is an enhancement of our powers of imagination. The data is there (tripling of plastic production by 2050), but the realization of the full implications of such an increase requires an imaginative power that the twenty-first-century brain must strive harder to achieve. As it stands now, statements and their full implications are out of sync. What is a critical impediment toward greater imaginative clarity?

As with so many of the current conflicts discussed in this book, *the haves* are not willing to share even a modicum of their resources with the *have-nots*, but want even more. What's also lacking is any appreciation that no one can view the problems from the outside looking in. The observers are themselves part of the problem—life is not a field trip. We are all ensnared in this plastic problem at varying levels of involvement.

(5) Only a fraction of the size of a human hair

Take a deep breath. The purity of the air you've just inhaled depends on many factors outside of your control. During any of the last three summers (2022, 2023, and 2024) the quality index of the air you just inhaled was affected by factors including the seemingly endless forest fires from the western United States and Canada.

Wildfires have doubled in incidence during the past twenty years. What's more, we seem to be experiencing the worst of it recently. The six most extreme fires have occurred since 2017 with the hottest year ever recorded at the time in 2023, which also marked the most extreme year for wildfires ever. Indeed, six of the most extreme fires have occurred since 2017.

While hundred-thousand-acre fires were rare two decades ago, they are increasingly common now and almost certainly will increase further. Assessing the burn rate of forests is like plotting the vertical trajectory of a military-grade rocket.

Neuroscientists now believe that wildfire smoke is worse for brain health—increasing the likelihood of dementia—than any other source of air pollution. The culprit is a fine particle *thirty times smaller* than the diameter of a human hair. These PM 2.5 particles (with PM standing for "particulate matter") readily enter the bloodstream when breathed in and are quickly ferried to the brain.

A research team funded by the National Institutes of Health (NIH) compared exposure to PM 2.5 from wildfires to other sources as causes related to new cases of dementia. To accomplish this, they looked into electronic health records of more than 1.2 million Kaiser Permanente (an integrated managed care consortium) Southern California members age sixty or older. To be accepted into the study, they had to be entirely free of any signs of dementia.

The researchers found that the risk of dementia rose between 18 and 21 percent for every one microgram rise of PM 2.5 particles. People from less-affluent communities also displayed a stronger association between wildfires, PM 2.5, and dementia. Overall the findings suggested PM 2.5 from wildfires causes a greater risk for dementia than PM 2.5 from any other sources. This finding may be related to the additional chemicals in wildfire smoke that are absent when the smoke stems from another origin. Or it may be related to the fact that low-income groups experience higher exposure to unhealthy air in general, coupled with a greater incidence of dementia, primarily Alzheimer's. So far the contribution of general air pollution to dementia hasn't been given the attention it deserves.

(6) "The Hot New Luxury Good for the Rich: Air"

Air pollution is continuing to expand like . . . well . . . a toxic cloud. Almost everyone in the United States breathes unhealthy levels of air pollution, according to the 2025 annual State of the Air Report published by the American Lung Association. Most concerning is particulate pollution (informally referred to as soot) consisting of a mix of solid and liquid droplets that float in the air in the form of dirt, dust, or smoke. Particulate particles are also created by coal- and natural gas–fired plants, as well as cars, agriculture, construction sites, and wildfires.

Only a tiny fraction of the size of a human hair, these particles bypass the body's usual defenses and become entrapped within lungs from which they pass into the bloodstream. The annual report states that short-term particle pollution, at its highest level in sixteen years, is most concentrated along the West Coast, with five of the ten most-polluted areas nationally found in California, with Washington and Oregon accounting for three of the other five slots.

So what can a person living in these air pollution hot spots do to lessen the harmful effects of pollution on the brain? Answers to such questions are already in the earliest stages of development. And, not surprisingly, the solution usually involves measures only the rich can afford.

If you were inside when you took your breath, the qualities of the air you breathed in varied with the money, time, and resources available to you. Only a few years ago many people believed (indeed some still do) that air quality equally affects the rich and the poor. Whether or not that was true in the past, it's no longer true in the new age of "climate gentrification."

In a groundbreaking article, "The Hot New Luxury Good for the Rich: Air," published in the *New Republic* on February 21, 2024, by journalist Sheila Love, air inequity was spelled out in sobering detail. Just as salaries, vacation destinations, luxury goods, executive expenditures, etc., have served in the past as class markers, now factors like the air we breathe are becoming a dominant determiner of value. One of two seemingly identical condominiums in the same building in the very near future may be twice the value of the other based on the quality of the air delivered and maintained in that unit.

This concept took off in 2020 when Gregory Malin, a San Francisco developer, started marketing the air inside of condominium residences as an amenity, just like a sauna, swimming pool, or garage parking place. Depending on the circumstances, the technology involves pulling air inward from the outside versus expelling indoor air outward. During COVID, the movement was forcing outside air inside in order to lessen air stagnation and the possible accumulation and distribution of airborne COVID inside. Just the opposite is now required in order to block air pollution from distant wildfires: sealing in the "good air," cleaning it, and preventing intake from outside air containing the small particulate matter from forest fires' smoke. This focus on air quality is not likely to be only a passing fad.

On June 7, 2023, New York City posted for a few hours the worst air quality in the world. "Hazardous," "unhealthy" air quality reports are no longer outliers but common descriptions of the orangish haze increasingly observable in cities affected by the extremes of pollution.

Elsewhere in the world of poor air quality "pay-to-breathe" economies are prospering. Already in China "lung-wash vacations," undertaken in converted bomb shelters, are all the rage.

Shayla Love summarized this "every man and woman for themselves" approach to air quality:

> The notion that smoke could be a democratizing force, afflicting everyone equally and perhaps motivating them

to take action to mitigate worsening climate conditions, is already colliding with the reality of an emerging luxury air market, yet another example of how, as the environment becomes less habitable, the wealthy will continue to insulate themselves from its worst aspects—even as their lifestyles disproportionally fuel emissions. As the fervor for ventilation that began during the pandemic meets the need to blockade against smoke, some wealthy people will do anything, and pay any amount, to guarantee they will always have a breath of fresh air.

Despite its importance, forest fire smoke, and the reaction to it—the selfish construction of personalized living solutions to a universal threat that affects everyone—hasn't received its deserved attention. Think of it as an example of "lifeboat mentality": In response to a potentially overwhelming phenomena like air pollution, people simply turn away from any universally applicable solution that would benefit everyone and, instead, turn inward. In response to their anxiety about the polluted air that is now becoming the norm, those who can afford it seek refuge in an expensive cocoon-like dwelling that, they trust, will save *them* however the rest of the world may fare.

Of all the health hazards threatening life on our planet, global warming tops the scale. Not only does it affect quality of life, but potentially over the long run can lead us on a slow walk toward Armageddon.

CHAPTER THREE

Global Warming

Thought Consequences from Global Warming

GLOBAL WARMING → Thought *

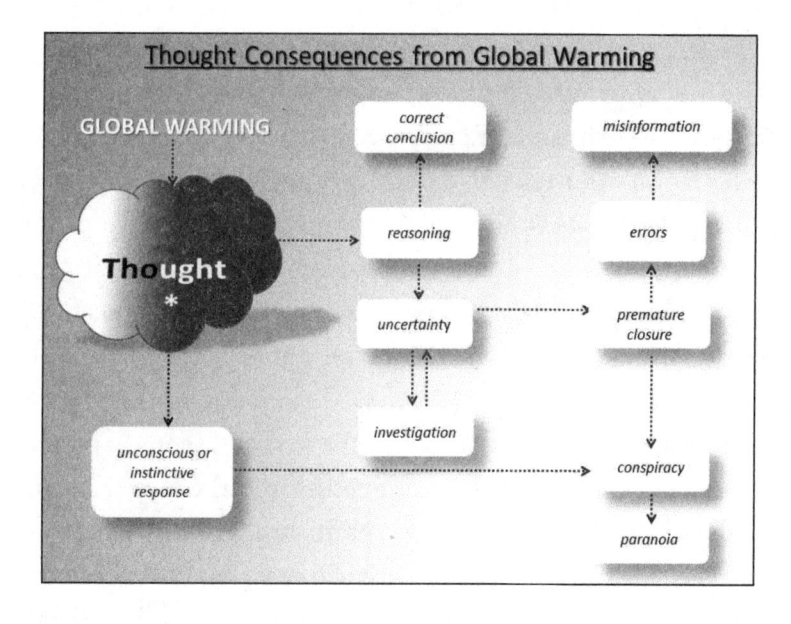

- correct conclusion ↔ reasoning ↔ uncertainty ↔ investigation
- misinformation ← errors ← premature closure ← conspiracy ← paranoia
- unconscious or instinctive response

(1) "Travel prepared to survive"

If you were in Death Valley in the summer and set off on a hike, you would soon encounter on the bulletin boards such advisories as "Heat Kills!" or "Don't become a Death Valley victim." At one of the final billboards you'd read the simple suggestion "Travel prepared to survive." For those intrepid walkers who ignore the bulletin's warnings and consequently plod onward in the extreme heat, you may later read about them in the papers or hear about their heat-related death on the local news.

We've all read or heard about the deaths of young and middle-aged people hiking in the desert or elsewhere during temperature spikes of 105 degrees or higher. How could this be? Why wouldn't a sensible person turn back when beginning to experience the early physical symptoms of heatstroke (nausea or vomiting, dizziness and/or weakness, alterations in sweating [excessive sweating or an absence of sweating altogether], rapid heartrate and breathing, and a throbbing headache)? Accompanying the physical symptoms are altered mental states such as confusion, the first indications of delirium, and agitation.

Recreation heat-related deaths are not only tragic, but also puzzling. Most of us would call it a day when encountering any of the symptoms of really excess temperature and return to our starting point. As a result of ignoring these warnings, many seemingly perfectly rational people when they set out on their hike . . . end up dead. What is going on?

Heat takes a toll on our reasoning powers and thought patterns.

In general the higher the temperature and the longer its duration, the greater the alteration in brain functioning. The brain's messengers (neurotransmitters) that carry signals from one neuron to the next undergo severe disruption. Neurons fire either too rapidly or too slowly, which leads to hugely destructive effects on thinking, memory, mood, and sleep. Compounding this is the loss of electrolytes through sweating—resulting in muscle cramps, weakness, and, in some cases, seizures. At the next level up, perception and behavior are both affected, especially at times requiring sustained attention, memory, and rapid decision-making—all leading to faulty performance.

Think back to occasions when you were immersed in a "heatwave" of moderate intensity by today's standards (92–100 degrees Fahrenheit). The most striking effect was your preoccupation with your discomfort. It was hard to think of anything else. You felt lethargic and listless. Concentration was nil, you couldn't "think straight"—a throwaway term for a lack of clarity and precision. Not surprisingly, under such circumstances productivity plummeted dramatically. Almost everyone experiences disruptions in mental clarity when exposed to high temperatures for at least a few days. If the humidity is also elevated, the cognitive impact is even greater.

If the temperature enters the three-digit zone of a hundred degrees or higher, oxygen levels are reduced, leading to rapid shallow breathing. Inflammation is spurred by the increased temperatures until, at some point, the body temperature spike begins to mirror the ambient temperature, with both increasing rapidly.

Next, blood flow to the brain increases. This is the body's last-ditch attempt to cool the brain and maintain the normal temperature. If this isn't successful, the brain swells, leading to disorientation, confusion, and eventually coma. Next, the so-called blood-brain barrier—the protective layer that protects the brain from potentially harmful substances in the blood—may be compromised, potentially allowing harmful substances to enter the brain and wreak havoc on its normal functioning.

If the temperature rises even higher, heatstroke results: a dramatic rise in the body's core temperature leading to confusion, stupor, coma, seizures, and, if the temperature is not promptly lowered, death.

(2) Heat and the frontal lobes

When it comes to lesser temperatures than those encountered in Death Valley, researchers have established how great the performance differences can be. Take the relationship between test taking and the environmental temperature.

Unless you are a closet masochist, you'd prefer to take an important examination in an air-conditioned room rather than a stuffy, warm one. No argument here. But more is at stake than just discomfort.

Scores on cognitive tests slump in tandem with temperature increases: A four-degree increase—expressed by the test takers as only mildly uncomfortable—leads to a 10 percent average decrease in performance on tests of memory, reaction time, and judgment. Other research found a .2 percent drop for every 1 degree increase above 72 degrees. Overall, the hotter the average daily temperature during a typical school year, the worse the students perform on standardized tests.

Most affected are those brain powers mediated by the frontal lobes: concentration, focus, and general executive function. Those factors are most vital in doing well on tests.

Most likely heat exposure over the continuum from mild to severe to extreme degrees selectively affects the frontal lobes of the brain, which are located behind the forehead.

When you sit in a chair with your elbow on a table with your hand cupping your forehead (sort of a "woe is me" gesture), the frontal lobes (one on each side) are behind the forehead bone you're pressing on your forehead.

Of all the areas of the brain, the frontal lobes are responsible for those activities that distinguish us from all other creatures. We are not alone in possessing frontal lobes, but in us the area is bigger, more developed than in other primates, and responsible for such functions as:

- <u>Sequencing</u>: keeping bits of information in sequential order. Related to this is separating the most essential information from less important background matters.
- <u>Executive Control</u>: planning or anticipating the consequences of one's behavior. Failures here can result in bad judgment resulting in catastrophes, such as serious car accidents resulting from driving faster than one's ability to control a speeding vehicle.
- <u>Self-Analysis</u>: loss of the ability to imaginatively project oneself into the future based on one's current actions.

Disorders of the frontal lobes lead to planning difficulties, poor decision-making, difficulty changing one's mind (perseveration, as neurologists refer to it), distractibility, and impulsivity.

Notice that these frontal lobe impairments under conditions of extreme heat provide an answer to our puzzlement about people willingly continuing to walk in extremely elevated temperatures. Going out for a hike under torrid conditions is the initial and ultimately fatal misjudgment followed by difficulty in changing one's mind and impulsivity. Add to these mental features the physical effects brought on by extreme heat, and you have a faulty action plan that too often ends in tragedy.

Less affected by heat are those areas responsible for arousal and warning; the sensory system that directs attention to the environment.

As a result you are aware that it's hot, and that you are perhaps slightly less alert. But the lion's share of your difficulty rises from your executive function problem in the frontal lobes.

Also affected is the anterior cingulate cortex, an area also toward the front of the brain associated with such mental processes such as detecting inconsistencies, errors, or unexpected correlations—the essential ingredients for successful test taking skills.

Extreme heat decreases the blood flow to the anterior cingulate, along with weakening of the functional connectivity between the frontal lobe areas vital for executive control. As described by Clayton Page Aldern in his book *The Weight of Nature*, "When people are exposed to high temperatures, the anterior cingulate unbuckles its activity from that of other cortical areas. The brain's fireworks become less coordinated and more randomized."

Let's focus on a few of the consequences of these heat-induced failures of thought and judgment:

- While driving, we precipitously turn in front of another car with just seconds to spare. Thanks to the heat we misestimated time and distance and therefore barely avoided an accident.
- While writing our rental check our mind wonders, and we enter an incorrect amount that represents the amount paid prior to the recent rental increase.
- As the temperature rises further, you experience difficulty thinking "straight" and remaining focused. This mini "brain fog" worsens as the temperature further increases.

Whatever the specific examples, the operating principle remains the same: The quality of our thinking is perpetually at the mercy of temperature.

Unless something is done to stop or at least moderate global warming, we may all soon be receiving the harsh lessons learned by the doomed hikers of Death Valley. But as we'll discuss later in this chapter, no solution for global warming is immediately in sight with the continuation of extremes of temperature almost guaranteed.

One more comment about Death Valley residents. In response to the life-threatening temperatures they are exposed to, many of the residents have already ditched their central air-conditioning as a result of spending thousands on electric bills. In such situations, cold water isn't available since the underground pipes overheat. If you turn on the cold water spigot, the temperature of the water will still be in the range of 108 degrees. The only way of gaining any relief is to turn off the hot water heater, thus converting the water tank into a reservoir that eventually cools to room temperature. But that water is still far from cool since the room temperature on many days exceeds 100 degrees Fahrenheit.

According to a prediction by the World Meteorological Organization, there is close to a 90 percent chance that a new heat record will be established over the next five years (2025–2030) as the earth's warmest year. There are simply no easy solutions or quick fixes for global warming. At least for now, it looks like we will continue to encounter what UN Secretary General Antonio Guterres refers to as, "The highway to climate hell."

(3) Lifeboat mentality

In September 2024, faced with the worst flooding in living memory, the European Union warned of a "climate breakdown." Floodwaters churned across Central Europe destroying homes and taking lives. "We face a Europe that is simultaneously flooding and burning. These extreme weather events . . . are now an almost annual occurrence," exclaimed Janez Lenarčič, EU crisis management commissioner.

According to a World Weather Attribution report, the ten deadliest extreme weather events in the last two decades have accounted for the deaths of more than half a million people around the world.

If you estimate the costs of repairing damage, emergencies, and rebuilding/recovery efforts, the figures are truly mindboggling.

- In the 1980s, disasters cost about eight billion euros per year.

- In 2021 and 2022, the cost topped fifty billion euros per year.
- The total EU costs, since the 1980s, exceeded six hundred and fifty billion euros.

We are reminded daily in the first quarter of the twenty-first century of the effects of global warming. By December 2023, the earth's average temperature was 1.5 degrees Celsius (2.7 degrees Fahrenheit) higher than average.

In 2024, nineteen thousand weather stations recorded higher temperatures starting on January 1. Over the ensuing ten months, each month exceeded the highest level reached in the same month in all previous years. In other words, the northern hemisphere experienced the warmest summer followed by the warmest winter in history. Further, the ten hottest years on record have all occurred in the last decade.

Four consecutive days in July 2023 were the hottest days in recorded history up until that date.

Not surprisingly, such meteoric increases in the frequency of national disasters have evoked a lifeboat mentality.

An example of this lifeboat mentality occurred in September 2024 at an international meeting held in Ghana by members of the United Nations. A series of talks and panels over two days aimed to establish ways of assuring fairness in the world's efforts toward reducing global warming. About thirty developing countries sent representatives. Despite the urgency, the United States, members of the European Union, and the United Kingdom did not send *any* representatives.

Within a week, a similar conflict between the haves and have-nots erupted at the two-day UN-sponsored climate-themed "Summit for the Future" held in New York City. In the face of "climate-fueled disasters," along with an increased debt burden affecting the poorest countries compared to the more technologically advanced countries, which distanced themselves from any obligation to help, the result, not surprisingly, was an increase in resentment among the poorest countries.

"International challenges are moving faster than our ability to solve them," commented UN Secretary General Antonio Guterres. "Crises are interacting and feeding off each other." Guterres cited as an example the spread of disinformation by digital technologies leading to a deepened distrust and an increase in polarization.

As Prime Minister Mia Mottley of Barbados phrased it, "The mistrust between the government and the governed, will continue to foster social alienation over the world."

Mistrust leads to denial, a powerful mechanism for avoiding truth. Denial is both the most common and the most difficult of the brain's defenses to break through. The negative consequences of relying on denial almost always exceed whatever the slight initial benefits may appear to be. China denied COVID until the evidence was overwhelming that a deadly disease was rapidly spreading worldwide. Similar denial took place here in the United States and other developing countries until the mounting death toll from COVID-19 eliminated denial with the rapidity of a smirk from a face at the receiving end of a slap.

On June 22, 2024, everyone I know was thinking about global warming, because of what so far has been the hottest day ever experienced in the United States nationwide. Some people's reaction was to hunker down in air-conditioned dens; others piled the kids into the car and headed for the beach several hours away; a third, a mentally unstable loner, shot three people and wounded seven as he opened fire outside the Mad Butcher grocery store in Fordyce, Arkansas.

(4) Heat-fueled aggression

Decades of research have established that as temperatures rise so too do riots, family violence, and aggravated assaults.

Heat aggression linkages can be traced as one moves "up" from simple organisms to primates. Monkeys launch more frequent attacks at one another once the temperature exceeds 80 degrees Fahrenheit (27 degrees Celsius).

When we reach the most evolved primate of all, us, heat-fueled aggression is sometimes hard to identify but it's there. For instance, the number and intensity of car honks in traffic has been experimentally shown to increase with higher temperatures. Heat-oppressed drivers telegraphed their mounting rage by spending more time leaning on their horns. When this cacophonous din reached a certain intensity, many of the remaining drivers responded in kind by honking their own horns.

During the longest sustained heatwave in the United States in decades, shootings took place in New York, Alabama, Missouri, and Ohio—resulting in a total of one dead and thirty-four others wounded. In the Alabama mass shooting, more than six hundred shots were fired.

As with all similar mass shootings, the usual explanatory tropes are evoked: too many guns, the US Supreme Court decision earlier in the month to strike down a federal ban on bump stocks (accessories that make it possible to fire a semiautomatic gun with the rapidity of a machine gun), etc. No mention is made, despite decades of confirmatory studies, of the firmly established linkage of temperature elevations and violence. This canary in a coal mine is not only ignored but throttled by widespread ignorance and denial.

Even at the highest levels of analysis, firearm violence is attributed to "socioeconomic, geographic and racial inequities," according to the US Surgeon General's 2024 report "Firearm

Violence: A Public Health Crisis in America." *Global warming and the increasing frequency of heat waves aren't even mentioned in the report.*

Of course we cannot prove the linkage of global warming and the specific rampage of the shooter in Arkansas. We are left to deal with probabilities, because that is the nature of the topics we are exploring. In the words of Rice University Professor Timothy Morton, "You would never directly experience global warming as such. Nowhere in the long list of catastrophic weather events—which will increase as global warming takes off—do we find global warming. But global warming is as real as this sentence."

(5) The Keeling Curve

Ever wonder why CO_2 is referred to as a greenhouse gas? Think of sunlight entering a greenhouse and warming everything up as a result of the accumulated heat, which is unable to escape because of the glass enclosure. A similar effect results from rising CO_2 levels in our atmosphere. The more CO_2 present in the atmosphere (acting as a glass dome), the warmer the climate becomes.

Early in the twentieth century, scientists speculated that the concentration of CO_2 in the atmosphere was rising secondary to fossil fuel consumption. But moving from speculation to fact was impossible until 1958 when Dr. Charles Keeling of the Scripps Institute of Oceanography developed a technique for measuring carbon dioxide concentration in the earth's atmosphere. He collected air samples at an observatory three thousand meters above sea level on the north side of Mauna Loa volcano on Hawaii's Big Island.

Today the eponymous Keeling Curve represents the world's longest continuous record of atmospheric carbon dioxide. Here is an illustration of the curve from 1960 to 2020.

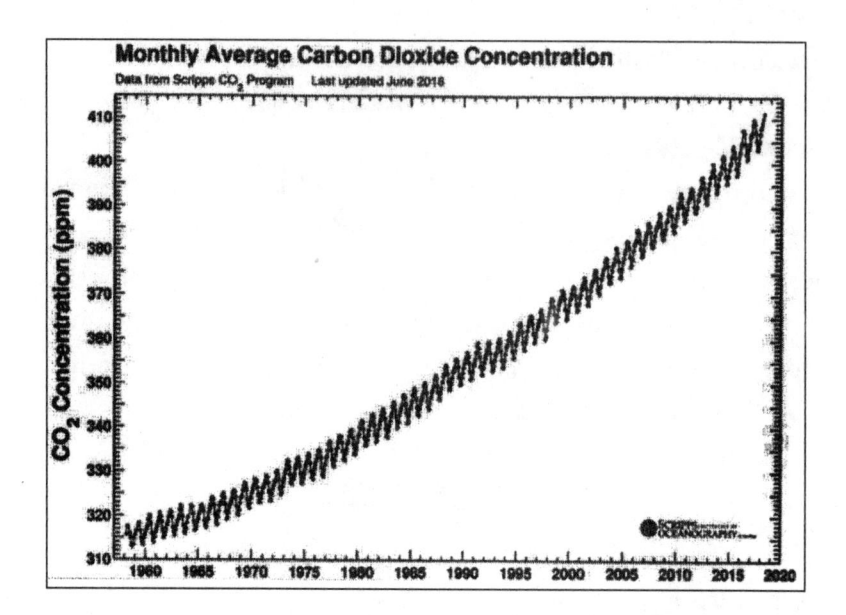

Monthly Average Carbon Dioxide Concentration
Data from Scripps CO₂ Program   Last updated June 2018

Note the inexorable drift upward. That curve tells us that in the sixty-two plus years since Keeling began his measurements, the amount of CO2 in the air we breathe has risen 33 percent, from 313 parts per million (ppm) to 411 ppm.

Notice the upward sawtooth pattern in the Keeling Curve. This represents a large seasonal pattern and reflects vegetation cycles in the northern hemisphere: CO2 is taken out of the atmosphere by living plants during the April–September growing season and released by dead plants during the rest of the year. Global warming results from failure to reach a balance between absorbing CO2 and releasing it.

Not only do we know the CO2 levels that prevailed during every year from 1960 through 2024, but we can compare these levels to hundreds of thousands of years earlier. How can we do this?

In the 1990s, enterprising scientists drilled deep holes in the ancient ice fields of Antarctica. Thanks to the ice cores they removed, we have an 800,000-year record of atmospheric CO2 levels embracing both ice ages and warmer spells. During these periods, the CO_2 level was never above 300 ppm (parts per million).

Throughout preindustrial times, the level varied between 275 and 285 ppm. But everything changed in the nineteenth century with the industrial revolution. By the early twentieth century, the level reached 295 ppm with a steady climb to 310 ppm by the 1950s. Since then the level has been ever upward: 325 ppm in 1970, 350 ppm in 1958, and 375 ppm in 2004. By 2015, the level for the first time in recorded history reached 400 ppm.

Not only is the level continuously rising, but it is rising rapidly. In the late 1960s, CO_2 levels rose at a rate of 1 ppm per year. By the twenty-first century, the rise reached 2 ppm per year with a 2.5 ppm annual measurement during the last ten years.

Rising CO_2 levels with each passing year are partly due to the tenacity of CO_2. Once in the atmosphere it can remain for hundreds of years. This is even more alarming when we consider that the gas continues to build up at an accelerating rate that so far outpaces our best efforts to control it.

Look again at the graph of carbon dioxide and its curve stretching upward, starting in the 1960s until the spring of 2025 when President Trump threatened to eliminate the National Oceanic and Atmospheric Administration, which provides some of the measurements contributing to the present and future progression of the Keeling Curve.

Global warming cannot be directly seen but, as with the Keeling Curve, it can be computed and graphed. True, the manifestations of global warming can be experienced in the form of weather such as temperature elevations and the consequent hurricanes, floods, etc. But weather and climate are not the same. While weather can only be reliably experienced and predicted for about one week (the "seven-day forecast" we hear about from television and radio meteorologists), climate change is measured in decades. Failure to appreciate this distinction between weather and climate accounts for such ludicrous assertions as "Yesterday we had three feet of snow. So think about that when you talk to me about global warming."

Notice as well that the Keeling Curve never experienced even an occasional dip downward or even any perceptible level of flattening. If this trend continues—and unfortunately there isn't any reason to believe that it won't—CO_2 will exceed 1000 ppm by the end of the twenty-first century. Here are some of the scary implications of reaching that 1000 ppm:

As the concentration of carbon dioxide rises, it exerts serious effects on our brain's ability to think clearly. "Carbon dioxide clouds the mind: It directly degrades our ability to think clearly, and we are walling it into our places of education and pumping it into the atmosphere" wrote James Bridle, in his prescient book *New Dark Age*. "The crisis of global warming is a crisis of the mind, a crisis of thought, a crisis in our ability to think another way."

Most jarring of all, most of the predicted twentieth-first-century threats are already upon us: "The future is already here—it's just not very evenly distributed," according to a quote attributed to science fiction writer William Gibson.

Apropos of Gibson's quote, check out the Weather Channel on any given day for a litany of thunderstorms, floods, tornadoes, rockslides, mudslides, hailstorms, airline turbulence, and, of course, record-breaking heatwaves.

We encounter here one of many self-referential twenty-first-century paradoxes:

In order to solve global warming and other existential threats, we have to be able to think clearly and logically; yet the very problems we are trying to address exert hugely negative effects on our brain's cognitive powers. Like a fish swimming in polluted waters, we can't extract ourselves from the medium in which we are immersed. Absent the ability to do this, we may soon lack the lucidity and cognitive vigor demanded for the task. As an even more sobering possibility, as Birdle put it, "Soon we shall not be able to think at all."

(6) Everything all at once

The twenty-first-century brain is faced with reconciling complications and contradictions about global warming, such as the following:

Initial erroneous assumptions greatly interfere with our understanding of global warming. Although CO_2 concentrations vary from one location and date to another (but it's always increasing as we know from the Keeling Curve), some of the largest contributions come from *stable* structures, such as cement buildings.

According to the American Institute of Architecture, about 40 percent of greenhouse gas emissions worldwide can be traced to the building industry, with carbon making up more than 10 percent of global annual emissions. Ongoing research is dedicated to developing concrete mixes that contain 30 percent less carbon dioxide than traditional concrete. One example of this approach is the Populus Hotel in Denver.

According to the claims of the hotel designer, Studio Gang, the Populus is the first "carbon positive" hotel: It sequesters more carbon than it emits. Although a step in the right direction, the Populus hasn't so far inspired similar approaches throughout the rest of the nation. Why? Because when we think of global warming we tend to consider gaseous sources rather than stable structures. It takes a bit of imagination to envision the solid bulwark of a building as continuously giving off harmful vapors that contribute to global warming.

While carbon dioxide (CO_2) is the primary greenhouse gas causing global warming, another greenhouse gas rarely spoken about is in some ways more potent and is warming the planet three times faster and ninety times stronger than carbon dioxide. I'm referring here to methane.

Concentrations of methane now are 2.6-fold higher than in preindustrial times. Despite its pivotal importance in spurring global warming, the gas flew under the chemical detection radar until only recently. One of the reasons for this stems from how difficult methane is to detect and track. Scientists now can identify methane using satellite and other remote sensing methods.

A major source of methane is the world's one billion-plus cattle, which belch more methane annually than the entire oil and gas industry, while leaks from gas stoves produce as much methane pollution as a half million US cars. It is considered likely that methane emissions will increase further from Arctic permafrost and tropical wetlands.

The third most important contributor to human-caused methane production comes from the organic waste in landfills and wastewater. Due to its propensity for trapping heat in the atmosphere (eighty times more harmful than CO_2 in the first twenty years after its release) methane is responsible for one third of total global warming since the industrial revolution.

Sulfur dioxide (SO_2) is another influence on global warming that may be even more important. Rather than just another contributor to global warming, sulfur dioxide may conceivably prove to be one of the solutions. While the more CO_2 released into the atmosphere, the higher the resulting temperatures, the more SO_2 released the greater the chances of a cooling effect, at least in the short term.

That's because the sulfur dioxide that forms as a by-product in the burning of sulfur-containing fuels (coal, petroleum, oil) leads to chemical reactions that release sulfur and water vapors. The resulting aerosols reflect sunlight *away from* the earth and back into space.

So why not simply take measures to increase levels of SO_2 while lowering the CO_2? Instead just the opposite has occurred: Sulfur has been reduced rather than increased. On January 1, 2020, the International Maritime Organization introduced new standards for shipping vessels aimed at cutting fuel sulfur content and resulting sulfur emissions by 80 percent. Thanks to this 80 percent reduction in SO_2 emissions, some scientists now believe the reduction was at least partially responsible for the stifling heatwave of 2024—at that time the hottest year in the previous two thousand years, claiming at least 2,300 lives, the highest number of heat-related deaths in over forty-five years of recorded data in the United States.

Global warming (some people still prefer the less informative term *climate change*) can't be avoided. We are constantly exposed to news stories about its effects. If we are watching the Weather Channel reporting about a heatwave in India that killed fifty-six people among twenty-five thousand heatstroke victims between March to May 2024, we can switch channels but we might well end up watching video feeds from other parts of the world featuring such catastrophes as:

<u>Nationally</u>: Tornadoes in Oklahoma and Kansas; torrential downpours followed by epic flooding in Texas and Louisiana; triple-digit heatwaves in Nevada and California; and wildfires in Hawaii, the Western United States, and Canada.

<u>Internationally</u>: 125-degree heat spells in India, China, and Iran; earthquakes in Turkey and Greece; volcanic eruptions or earthquakes in the Pacific Ocean followed by tsunamis. Those of us lucky enough to never have personally encountered such extreme events can only helplessly watch via the seemingly intimate but actually psychologically distancing perspective of our iPhones and iPads. (More about that in the chapters on the internet and AI)

CATASTROPHIC WEATHER EVENTS

As the temperature rises, the incidence increases of catastrophic upheavals affecting land, water, and wind. The acceleration of *wind* speed produces hurricanes on water and fire on land, as happened in January 2025 in Los Angeles. When the wind speed reached high double digits (70 mph and above), dry combustible plant material resulting from the previous months of drought led to the spread of small fires that led to a refulgence that over three days devoured over thirty thousand acres.

Water-caused catastrophes vary according to the amount of water involved. When rain reduces to a trickle, drought conditions provide the tinder for igniting the seemingly unstoppable rings of fire that encircle whole communities. Too much water and the end results are torrential rains leading to flooding, landslides, and home destruction with associated loss of life.

Land catastrophes from high temperatures result in the dying of plants and their desiccation into underbrush, which is exquisitely sensitive to even a spark of fire.

A carelessly thrown cigarette can bring down a forest.

The force field vector diagram on page 41 illustrates the human consequences of water, land, and air catastrophes that lead to associated neuropsychiatric afflictions, which can include PTSD, mourning, depression, lingering anxiety, loneliness, familial conflict, homicide, suicide, spousal and child abuse, and helplessness.

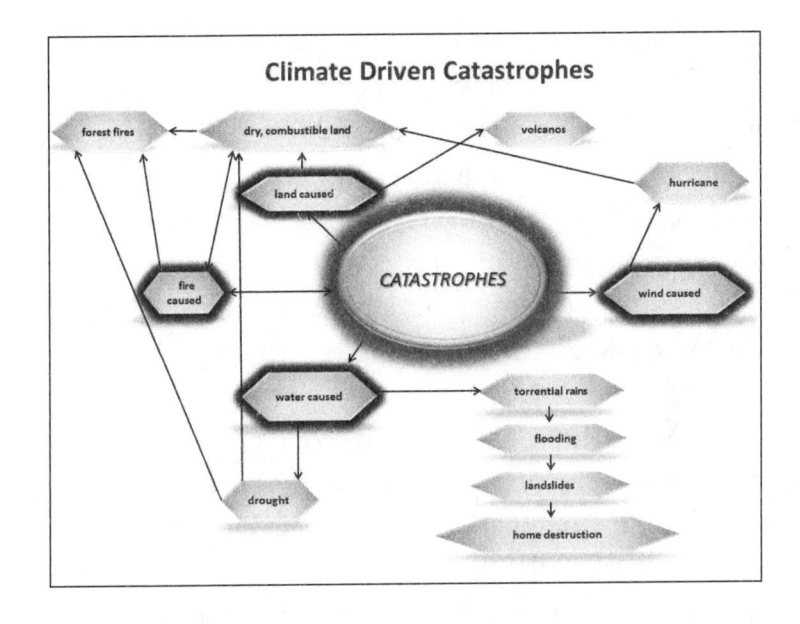

Global Warming

Climate Driven Catastrophes

- forest fires
- dry, combustible land
- volcanos
- land caused
- hurricane
- fire caused
- CATASTROPHES
- wind caused
- water caused
- torrential rains
- flooding
- landslides
- home destruction
- drought

Weather Events

- Loss of life or physical impairment
  - no taxes collected from people affected
- destruction of property
  - huge government cost FEMA
    - increases in national debt
- a new homelessness problem involving middle class
  - stress related illnesses
  - mental illness rates soar
    - PTSD
    - anxiety disorders

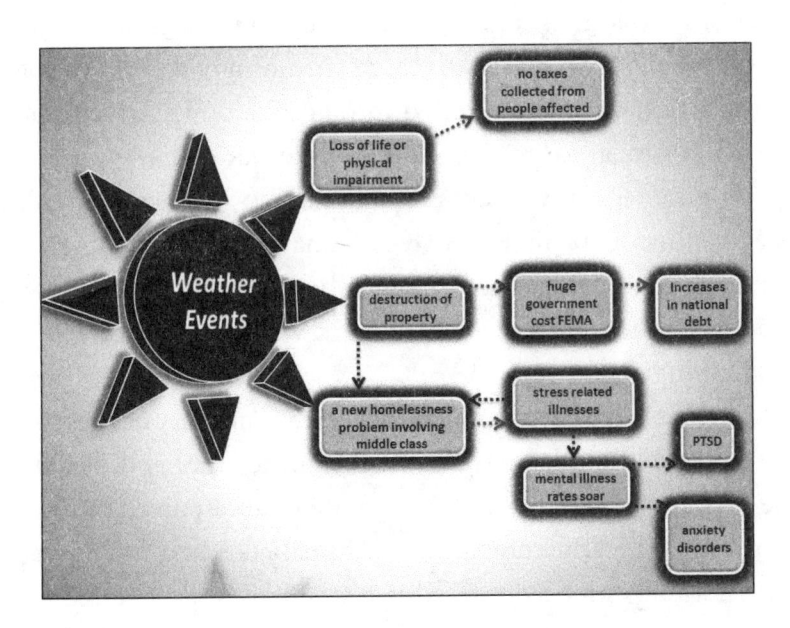

Social and cultural tsunamis have also increased in frequency and intensity. Almost daily we encounter on the nightly news mass shootings displayed in vivid detail: children killing their parents or parents killing their children; widespread, often violent demonstrations such as those over the Israeli-Palestinian war, which paralyzed some of our major universities in the spring of 2024 and into 2025; and seemingly inexplicable attacks with knives or machetes directed at randomly chosen strangers encountered on our city streets.

In tandem with global warming, all of these tumultuous events are now occurring simultaneously. Our brains attempt to parse out patterns and meanings from these interrelated disasters. But since only a small percentage of us directly experience such disasters at any given time, few of us feel any sense of urgency. After all, a hurricane or a wildfire or a knife attack on the street are occurrences that most of us have only seen on television.

But we no longer have to watch streaming videos from India of heat-stricken victims wheeled into hospitals. We can currently encounter similar horrors in some parts of our country where the temperature regularly reaches triple digits (105 degrees Fahrenheit in early June 2024 in Phoenix, Arizona). While watching such scenarios, we experience an initial disbelief followed by a sense of hopelessness and anxiety consequent to viewing a horror we can't do anything about.

After we have been exposed to a certain number of these anxiety- and dread-inducing scenes (the requisite number differs from one person to another), our brain's emotional responses transform from sympathy and compassion to impatience and finally anger that "nothing can be done." This response is aptly referred to as *compassion fatigue*: Our cache of concern has slowly dwindled to a trickle, and it's difficult to empathize with the people so graphically depicted. None of us want to be reminded of our inability to help the sufferers. As a result, after a certain number of exposures we are forced into MEGO syndrome (My Eyes Glaze Over), and we turn away. As one mother interviewed by the *New York Times* stated after watching images of a climate-induced famine: "I get upset watching the babies dying. Who the hell wants to see that? I switch the channel."

As bad as things are, we may be underestimating how bad things can actually get. For instance the amount of carbon monoxide in the atmosphere may be increasing even more than we think. In 2024, the rate was 27 percent larger than the previous record set in 2015—skyrocketing the total atmospheric carbon dioxide concentration to a level not seen for three million years. The reason for this increase is likely due to several factors: forest fires or logging operations releasing centuries of accumulated carbon over the duration of a single day; prolonged droughts terminating plant photosynthesis and interfering with carbon uptake; and ever-rising temperatures leading to greater soil decomposition, releasing additional carbon dioxide back into the air. As the forests and plants disappear, we are, to that extent, deprived of our "sinks" that absorb carbon dioxide. This change can only worsen the situation.

It's been estimated by NASA's Orbiting Carbon Observatory, which provides measures of carbon dioxide from space, that landmass absorbed close to 2.24 fewer gigatons of carbon dioxide than expected over a twelve-month period of measurement—roughly the equivalent of burning nine billion pounds of coal. According to Sarah Kaplan, a reporter for the *Washington Post*: "The land and oceans have historically taken up about half of the greenhouse gases people emit. Without this absorption of CO_2—known as carbon sinks—global temperature rise would be twice the roughly 1.3 degrees Celsius (2.3 Fahrenheit) the world has endured."

In other words, we may pay a high price for the slow walk we are taking in our efforts to reverse, or at least moderate, global warming. In regard to the unexpected potential climate disasters, we may not be clearly grasping what is happening, much less responding to and controlling it. Hidden, unexpected surprises may await us like a final Halloween skeleton that, at the last moment, drops and dangles in front of our face as we exit the haunted house.

One thing is for sure: The pace of global warming is picking up at a faster rate than most scientists believed possible. The year 2023 reached 1.5 degrees Celsius of warming (about 2.7 degrees Fahrenheit) over the late-nineteenth-century baseline at the start of the industrial revolution. Things worsened considerably in 2024 when we reached 1.6 degrees Celsius.

If you read the official announcements by various panels and committees, global warming is described in less urgent terms than when the same scientists during individual interviews are asked their opinions "off the record." This distinction between communal versus individual opinion results from how scientists manage uncertainty. Physical science research adheres to the scientific method based on experiments and observations; conclusions can be evaluated by other scientists ("peer review"), and the results must be shown to be reproducible by other scientists. When a specific scientist is asked for his own expert opinion of "What's going on, but I can't prove," many scientists remain silent rather than speak out in the face of uncertainty. One study "found scientists 'conservative' in their projections of the impacts of climate change" with a tendency to "err on the side of least drama" along with overt "adherence to the scientific norms of restraint, objectivity, skepticism, rationality, dispassion and moderation." Such reticence can be dangerous, according to America's most well-known climate scientist, James Hansen of Columbia University: "Scientific reticence hinders communication with the public," he observed.

We are currently witnessing all around us climate catastrophes that were "officially" declared unlikely to begin until the dawn of the twenty-second century. As a result, it's difficult to draw up a valid course of action now based on mistaken conclusions that are, essentially, a form of misinformation. Permafrost is a stellar example: "Permafrost is thawing much more quickly than models have predicted, with unknown consequences for greenhouse-gas release," according to climate scientist Merritt Turetsky at the University of Colorado Boulder.

On top of all this we now have an administration in Washington, DC, that—when it comes to climate change—employs the most primitive mental defense mechanism of all: denial. For example, the 2025 Annual Threat Assessment by the US Intelligence Community made no mention of climate change for the first time in eleven years! This dovetails perfectly with the scientific reticence discussed in detail in the March 25, 2025, *Bulletin of the Atomic Scientists* by contributor David Spratt, currently research director of the Breakthrough National Centres for Climate Restoration, and author of the 2008 book *Climate Code Red: The Case for Emergency Action*.

With the planet changing faster than anticipated, underestimation of risks "inhibits public understanding and policy making," wrote Spratt. "Knowledge is relatively new and growing quickly; the physical reality is changing rapidly, uncertainty can become an excuse for political delay; and the cost of not understanding the systemic risks may be the viability of major earth systems and human societies."

But things don't have to be this dark. There is a precedent we can point to that offers hope.

(7) A triumph

Scientists can point to one environmental problem that was nipped in the bud: ozone. Without a viable ozone layer, we'd have no protection from the harmful cancer-causing ultraviolet radiation emanating from the sun.

Signed in 1987 by nearly two hundred countries, the Montreal Protocol aimed at protecting the earth's ozone layer by steadily decreasing the production and consumption of over a hundred ozone-depleting substances (ODS): man-made chemicals that damage the ozone layer.

Thanks to the Montreal Protocol, ozone-destroying chemicals were phased out in what scientists agree was the most powerful and successful environmental treaty ever written.

Signatories to this treaty agreed to take on specific responsibilities related to the eventual elimination of ODS. The biggest target was hydrochlorofluorocarbons (HCFCs), which are gases used throughout the world in refrigeration and air-conditioning.

At the time it was known that HCFCs are found in both ODS and greenhouse gases. But what wasn't appreciated was the powerful effect of HCFCs on global warming: two thousand times more potent than carbon dioxide.

Serendipitously, the Montreal Protocol, by reducing HCFC levels, provided an unanticipated benefit on global warming. Scientists estimate that the treaty has averted an additional one degree Fahrenheit of global warming by 2100.

"In trying to stave off one planetary disaster, the signatories bought us some time on another one," observed Jonathan Mingle in the *New York Review of Books*. When it comes to preserving the ozone layer there are no losers, only winners.

Why was the Montreal Protocol so successful? Why does nothing like it exist, at least for the moment, for global warming?

Possible explanations include:

1. Preserving the ozone layer was universally recognized in 1987 as a benefit well worth the cooperative efforts required

to achieve it. No such sense of unanimity exits in our 2025 "every man for himself" fractionated world.
2. Global warming is currently a pocketbook issue with different groups engaged in fiercely competitive self-interested efforts to *expand* or drastically *curtail* the continuation and expansion of fossil fuel uses. With money playing an inordinate role here, greed has become so motivating a factor that progress has come to a halt.
3. This dualism between expanding and curtailing the use of fossil fuels is no doubt a reflection of the brain's inherent dualism: The brain is divided into two separate, though interconnected, hemispheres, each with its own specialized function and interests.

"Men are divided in opinion as to the facts. And even, granting the facts, they explain them in different ways," wrote Edwin Abbott, the author of *Flatland*, the Victorian scientific fantasy of imaginary beings, Flatlanders, who lived in a two-dimensional universe consisting of a flat plane and who possessed no knowledge of a third dimension.

When it comes to global warming and the other challenges facing us, many of us remain Flatlanders, whose experiences are confined to one plane with no awareness of anything outside that plane. Thoroughly immersed in Flatland, we seemingly cannot rise above the surface and envision an enriched alternative to the flat surface.

(8) Unanticipated consequences

"Technical problems will always be solved by means of technical solutions, if you wait long enough."

That aphorism doesn't carry anybody's name affixed to it; no doubt because it seems too obvious to merit any formal citation.

But many of the technical problems that the twenty-first-century brain is called upon to solve don't always respond to technical solutions. But let's start with an example of a successful technical solution for a technical problem, namely food storage.

Temperature-associated food spoilage was the third leading cause of death at the start of the twentieth century. Within three decades after the availability of refrigeration, the incidence of death by food-associated infection and inflammation fell by about 85 percent. As a consequence, human behavior was forever altered: weekly rather than daily shopping and food transportation nationally and internationally, along with local distribution worldwide. Entire ecosystems took shape as customers started taking for granted the year-round availability of rice from India, peppers from Mexico, and meat from Australia and New Zealand.

But a technological solution like refrigeration often carries—in fact usually does—a lot of luggage crammed with dirty laundry. The cold inside your refrigerator is not a force in itself, but the absence of heat that has been shunted into the surroundings.

Thus the refrigeration process creates more heat that further contributes to global warming. Another liability of refrigeration stems from its effects on our diet. Meat eating was sustained by the widespread availability of refrigeration from the 1930s on. Even though the American prairie with its mega herds of cattle is long gone, nobody today who is hungry for a hamburger experiences any difficulty in finding one.

In contrast to successes like refrigeration, global warming hasn't yielded to any scientific solution. While almost everyone accepts that the rising CO_2 levels help explain global warming, all of the proposed solutions for lowering it up until now have involved mostly behavioral abjurations: Replace your gas-guzzler with an electric car or refrain from doing anything that requires the consumption of fossil fuels, etc.

One proposed solution involving sulfur carries a hefty dollop of unintended consequences. It was suggested in 2021 by science fiction writer Neal Stephenson in his novel *Termination Shock*. The character billionaire T. R. Schmidt constructs a massive gun that shoots sulfur in the form of SO_2 into the atmosphere in order to reduce global warming. Schmidt's scheme is based on an observed linkage between a volcanic eruption in the Philippines in 1991, which sent a cloud of volcanic ash into space. It's estimated that seventeen million tons of sulfur dioxide resulted from the eruption of Mount Pinatubo and spread across the stratosphere. As a result, much of the sun's rays were reflected away from the earth, which led to a drop in average temperature in the northern hemisphere by almost 1 degree Fahrenheit. This was maintained over the next few years.

Today, the concept of cooling the planet by means of sulfur dioxide infusions into the upper hemisphere is no longer a plot element in a science fiction novel. It's been featured in a lengthy front-page article in the *New York Times*. "Scientist Wants to Block Sunlight to Cool Earth" is one of a series of articles devoted to altering the stratosphere in response to global warming. The article discusses many of the objections some scientists hold toward saturating the stratosphere with sulfur. For instance, Stephenson's title *Termination Shock* refers to the possibility of deadly consequences for the planet if the system of sulfur saturation of the stratosphere works as planned, but at some point in the future the system needs to be shut off.

But the threat of possibly deadly consequences if the sulfur saturation has to be abruptly reduced isn't currently stopping venture capitalists from releasing modest amounts of sulfur into the stratosphere. One of them, Make Sunsets, a start-up company based near Saratoga, California, was profiled in the *New York Times* in late September 2024. The article detailed how this small company is attempting to affect sulfur dioxide levels on a tiny scale.

The process starts on an isolated patch of dirt on an undeveloped hillside owned by one of the cofounders of Make Sunsets. After partially filling a balloon, with about 1.7 kilometers (almost four pounds) of sulfur dioxide, the balloon is topped up with enough helium to completely fill the balloon. At that point the balloon is sealed with electrical tape and a GPS tracker is attached. After the balloon is launched, it travels on average about fifty miles, reaching an altitude of thirty-five kilometers before the buildup of atmosphere pressure pops the balloon, dispersing a minuscule amount of sulfur dioxide into the stratosphere.

While I wouldn't describe such efforts as ludicrous—I'm not a geoengineer—one person who knows a lot about the science had this to say to a *New York Times* reporter: "They are a couple of tech pros, who have no experience in what they are claiming they do" according to Sikina Jinnah, a professor of environmental studies at the University of California, Santa Cruz, and a recognized expert in geoengineering "They are not scientists and they are making claims that no one has validated."

Two things stand out in all of this. First, the rather fringy idea that barely measurable amounts of sulfur dioxide (certainly in comparison with the stratosphere in which the chemical is dispersed) can make a measurable difference in the amount of sunlight deflected back to the sun. Second, Makes Sunsets, along with other start-ups to come, are very much for-profit enterprises (the participants would prefer the term entrepreneurial enterprises). You too can be part of the Make Sunsets gig by purchasing "cooling credits" currently selling at $2,200 a credit. There is only one thing missing so far in this zany scenario: a juicy conspiracy theory.

So far little progress has been achieved toward altering the likely progression of climate change. True, money is pouring into efforts to control, or at least modify, extreme weather events caused by climate change: floods, record-breaking heatwaves, forest fires etc. So far the methods suggested (reducing fossil fuel use, along with more drastic measures aimed at coming up with technologies to lower atmospheric carbon dioxide) have had no track record or a bad track record of success. You've only to turn on the nightly news and watch the forest fires, flooding, and hurricanes to see how these approaches are faring.

But whatever relief may come from these preventive measures, they also come at an unacceptable cost: The longer we wait to address global warming, the more damage will be done and more money will be spent to fight the impact of even hotter weather.

Some experts are predicting that the ultimate solution for the climate problem will come from AI. AI will be discussed in more detail in chapter 6, but for now let's take a short overview.

(9) A David Lynch–inspired landscape

Using AI as a solution for global warming involves a paradox: The very thing we are using to lower CO_2 levels and relieve global temperature elevations is itself responsible for exacerbating the problem.

According to Kate Crawford at the University of Southern California: "We are building infrastructure for artificial intelligence that is extremely energy and water intensive, without looking at the very real downsides in terms of the climate impacts. Particularly concerning are all the large scale data centers which employ enormously heat producing GPUs [graphic processing units]. And the water to cool the GPU chips is fresh water, which often comes from the same reserves that we use for drinking water."

Data centers in Silicon Valley alone contain over 300,000 square feet of servers, many of them operating 24/7. Worldwide, the global energy budget as of last year for one data processor (Equinix with its nearly eleven thousand centers worldwide) resulted in a global energy demand similar to the energy requirements of three-quarters of a million US homes. What's more, the global energy use at the world's current data centers has been estimated as somewhere between 2 and 8 percent.

Today a data center complex in Iowa owned and operated by Meta uses as much power per year as seven million laptops running eight hours a day. Google announced in July 2024 that it was well behind on its pledge to eliminate its net carbon emissions by 2030. Its emissions have already increased nearly 50 percent since 2019.

If you live on the East Coast you can personally observe a similar array of wall-to-wall servers along sections of the highway leading from Washington, DC, to Middleburg, Virginia. Drive that short distance and you'll experience a David Lynch–inspired landscape of huge unnamed and unidentified buildings with opaque windows surrounded by high wire fences and metal barriers with no sign of life anywhere.

With new data centers coming online every week or so, by 2027, 8.5 percent of worldwide energy use for AI will equal the electricity requirements for Argentina over a full year. By 2030, US electricity demands might be as high as 20 percent of all the electricity produced.

As *New York Times* science writer David Gelles summarized, after a thorough review of the current thinking about AI and its energy requirements, "When the tech companies themselves are consuming all that electricity towards powering new AI data centers, pushing up energy demand, it isn't making the grid overall any cleaner. AI may yet deliver breakthroughs that help reduce emissions. But, at least for now, the data centers are doing more harm than good for the climate."

In the meantime, we are promised, without any solid data to back it up, that additional and advanced AI will solve the energy use problem, and the resulting gains in efficiency will make up for any additional energy demands. At this point, such a view has more in common with a religious belief than a scientific certainty.

And all of this comes in response to grim forebodings that nations around the world can't or won't get global warming under control.

(10) "Only a rational business decision"

During the autumn of 2024, optimism about controlling climate change was starting to wane within the industries that really matter: the gas and oil companies.

Among companies that embraced wind or solar technologies, or electrical vehicle charging stations, their stock values plummeted. BP, an early adaptor of alternative energy, after pledging in 2020 to reduce its oil and gas production by 40 percent by the end of the decade, reversed course and increased its spending on fossil fuels and sold off its wind assets. During that period its stock suffered a loss of 19 percent of its value.

In contrast, among those energy companies that resisted attempts to persuade them to adapt renewable energy and remained invested in fossil fuels, stock prices rocketed to a profit of more than 70 percent since 2019.

The forces at work here come as no surprise. Whether oil and gas companies stick with fossil fuels or reduce their carbon footprint depends on which course of action results in the greatest profit. "Some people jumped the gun and moved towards some paths that ultimately turned out to be, I'd say, devastating to their bottom line," Toby Rice, the chief executive of the Pittsburgh natural gas producer, EQT, told *New York Times* reporter Rebecca F. Elliott.

As a result of financial losses, business is less inclined to make the necessary sacrifices to reduce greenhouse gases. For instance, Ford Motors and General Motors have put the brakes on ambitious plans for new electric models in response to decreasing customer demands, while Volvo canceled entirely its initial goal of converting to electrical vehicles by 2030.

Another indication of a waning in confidence in climate control are the changes in approach to climate goals that US banks are taking. Morgan Stanley, JPMorganChase, and the Institute of International Finance have become convinced that the goal of limiting temperature increases to 1.5 degrees Celsius or lower is "almost certainly unachievable," according to a report by the Institute of International Finance. While the banking industry publicly continues to support a transition from fossil fuels to clean energy, their advice to their clients is based on a pessimistic view of how things are likely to go over the years from 2025 to 2030.

"We now expect a 3° Centigrade [rise]," Morgan Stanley analysts wrote in March 2025. Mega banks like Wells Fargo are hedging their bets on their earlier climate pledges and exiting from the Net-Zero Banking Alliance. Morgan Stanley's Client for Climate Forecast was hidden away within a research report on the future of air-conditioning stocks. "A 3° of warming could more than double the growth rate of the $230 plus billion air-conditioning market every year from 3% to 7% between now and 2030." As the authors of this forecast phrased it, "The political environment has changed," no doubt a reference to President's Trump's withdrawal from the 2015 global agreement as part of a long-term effort to dismantle environmental rules originally designed to limit global warming and environmental pollution.

"The political environment has changed. So some of the [banks] are conforming to that," according to Gautam Jain, a senior research scholar at Columbia University, who concludes, "But mostly it is a rational business decision. Betting on potentially catastrophic global warming is both an acknowledgement of the current emissions trajectory and a politically savvy move in the second Trump era," in an interview with Corbin Hiar of E & E News.

You might want to read that last paragraph again to get the full flavor of the greed and narcissism that all of us have to contend with in attempting to resolve global warming.

(11) The unwilling or unable

Unfortunately this pessimism by the gas and oil industries and the banks about decreasing global warming has entered mainstream American thought. Consider the increasingly passive acceptance of global warming in some circles. For instance, it's increasingly common on unduly hot days for women (and an increasing number of men) to sport battery-powered mini fans. At moments when the heat approaches unbearable levels, the tiny fans can be discreetly removed from a pocketbook or man bag, turned on, and directed toward the owner's face for as long as it takes for the oppressive sense of heat to pass. Increasing numbers of people are coming to the realization with each heatwave that they need to be prepared for a world that, if things continue along the lines of Gautam Jain's advice, is always going to get hotter and hotter.

In *Terminal Shock*, science fiction writer Neal Stephenson describes the next step beyond fans: The earthsuits that may await us in the future. "An earthsuit is not so much a single garment as a toolkit of parts that could be snapped together in different ways depending on conditions. This refrigeration system discharges heat back into the environment."

In this perhaps prophetic novel, the characters resort to earthsuits during the hottest parts of each day. Nor are such developments likely to be confined to science fiction. In real life even more advanced solutions to potentially deadly heat are already available.

IcePlates, made by Knoxville, Tennessee–based Qore Performance, fit into vests that are currently worn in industries ranging from fast food to warehouses—anywhere workers are at risk from heat injury or death.

The impetus for these vests can be traced to the United States Department of Defense. The original goal was to provide a cooling device capable of providing two hours of cooling without adding any weight to combat soldiers, who already labor under a heavy load. The army's requirements, it turned out, were initially impossible to achieve. So everybody went back to the drawing board until a former employer of the Naval Special Warfare Development Group (known in the military as SEAL Team Six) described to Justin Lee, chief executive of Qore, how during hot missions in Afghanistan and Iraq combat troops would freeze packs of drinking water and place them at strategic places beneath their body armor. With further tinkering, the concept of body cooling resulted in rectangular "plastic Plates" that fit the contours of the human body. This system holds fifty ounces of frozen water.

The use of "cooling vests" or "icepacks" is already an accepted practice in Formula 1 starting after the 2023 Qatar Grand Prix. By keeping the driver's core temperatures cool, the vests help combat fatigue and thereby enhance performance.

What lies in the future? Probably something akin to the "earthsuits" for all of us with built-in sensors able to identify those parts of the body that are most in need of cooling.

Despite disappointing approaches so far in combating climate change, we must overcome this twenty-first-century challenge and make innovative decisions about ways to save energy and reduce greenhouse gases. And at the same time, and equally important, we must overcome resistance from special interest groups, who are unwilling or unable to make the personal sacrifices that will be required. So far pitting these seemingly irreconcilable points of view against each other has only spurred antagonism, conflict, and disharmony. Even so the brain must press on while seeing both sides of the issue: While everyone wants the benefits of reduced fossil fuel use, this isn't going to happen as long as some people's economic well-being remains tethered to the status quo. The twenty-first-century brain must come up with some solution to this highly conflictual conundrum. "The goal is still worthy, but we have to think about the collateral damage of all of our major decisions," stated New York Governor Kathy Hochul at an autumn 2024 climate event.

Will the twenty-first-century brain rise to the occasion by devising ways of controlling global warming, or will our lives be marked by the reluctant acceptance, but acceptance nonetheless, of selfish and self-centered workarounds with mini fans and earth-suits followed by huge increases in air-conditioning and entirely climate-controlled homes as discussed earlier—for those who can afford it?

At this point, let's examine some helpful background about the brain, which will be helpful in clarifying the chapters to follow, especially those on the internet and AI.

CHAPTER FOUR

The Twenty-First-Century Brain

(1) Probabilistic conclusions

While it takes perhaps thousands of years for the human brain to undergo anatomical changes detectable only to the trained eye (a pathologist or neurosurgeon), changes in our understanding of the brain can take place in a decade or less.

Two principles are at work here: *localization* (specific brain regions dedicated to designated tasks) and *integration* (communication involving varying, sometimes vast, numbers of brain regions).

According to the thinking that held sway throughout the nineteenth and early to mid-twentieth centuries, each brain consisted of subdivisions responsible for various functions, i.e. speech centers, movement centers, hearing centers, etc. Underlying this was the belief that each of these centers is almost identical from one person to another. With further research it became clear that this understanding of brain organization was an oversimplification. In terms of speech, for instance, many other brain locations are important in addition to the classical speech areas. Localization— the guiding principle of pre-twentieth-century thinking about the brain—has given way to probabilistic rather than certain conclusions about location and function. We have gone from a "Neuron A is connected to neuron B" understanding to "Neuron A is *probably* or even *possibly* connected with neuron B."

At the microscopic level, neuroscientists in the late nineteenth century began showing how neurons are interconnected at synapses—tiny spaces separating one neuron from another. The word synapse is based on the Greek words "syn" (together) and "haptein" (to clasp). But instead of a material clasp, the synapse consists of an infinitely tiny space separating two nerve cells through which messages are passed from one to another. A single neuron within the cerebral hemisphere may have thousands of neurons attached to it via its synaptic connections. This arrangement is so tight that the average human brain contains about eighty-six billion neurons. If you extract a piece of the human cortex, approximately the size of a grain of rice, it contains about ten thousand neurons, with the whole brain making up the eighty-six billion neurons total. If each neuron creates synaptic connections with hundreds or even thousands of neurons, the total number of these communication points is in the trillions. Current estimates are around 0.15 quadrillion or 150,000,000,000,000 synapses.

Each of the eighty-six billion neurons includes a fiber called the *axon*, which carries an electric message to other cells, along with a varying number of branching fibers called dendrites that receive the axonal messages from other neurons. Information is initially conveyed through electrical signals from the axon of one neuron and ending on a receiving dendrite or multiple dendrites on the other neuron or neurons. At the *synapse* the electrical signal conveyed by the transmitting axon is converted into a code that leads to the release of chemicals (neurotransmitters). These messengers cross over the synaptic cleft to specialized receptors on the membrane of the second neuron.

The brain's functions are related to the number of neurons and their interconnections. Behavior is the culmination of the activity of many, many nerve cells linked together in neural networks. At this point, here are the key factors and concepts:

1. Each nerve cell is unique.
2. Function is dependent on transmission across the synapse of many different messenger chemicals (neurotransmitters).
3. The brain does not exist in isolation, but is involved in a two-way communication with the outside world.

Consider that third point for a moment. While it's true the brain can engage in thinking without any sensory input (nothing seen, heard, felt, smelled, or tasted) such as when a person sits in a darkened, soundless room, eventually the brain calls out for some kind of contact. To some extent, therefore, our brain when we are awake is always interacting with the outside world.

(2) A huge bowl of spaghetti

Analogically you can think of the brain as a huge bowl of spaghetti occupying a large container composed of three coverings: the pia mater, a thin diaphanous and transparent layer that makes direct contact with the brain; the dura mater ("hard mother" in Latin), which fits snugly over the brain like a hand enclosed in a correctly fitted leather glove; and finally the bony outer skull.

Now go ahead and cut (but just a bit), mix, and toss the spaghetti. Now grasp one strand and *very slowly* start removing it. Since there is no way of knowing ahead of time the length of the strand, you have no way of reliably estimating exactly how long it will take to remove it. Nor can you estimate beyond a wild guess the number of other strands that may be in contact with the particular strand you've chosen.

Consider those spaghetti strands as comprising a neuronal network that transmits messages via the *axons* and *dendrites*. These two components are responsible for the mindboggling number of neuronal connections. Because of these vast connections, the loss of one neuron with its axon and dendrites—one strand of the spaghetti in our analogy—is not likely to be any more significant to the brain's overall functioning than the decision by one voter to remain home on Election Day is likely to exert any detectable effect on national election results. The brain's power results from the network formed by the neurons, rather than the action of any singular neuron by itself.

Currently neuroscientists take a more dynamic view based on a dispersal of functions throughout the brain. The emphasis is now on variations both in time (milliseconds) and distance (millimeters) among the brain's connections. *Dynamic organization and reorganization is the key to our new understanding of the twenty-first-century brain.* This does not imply an "everything is connected to everything" oversimplification. But, as with the separate strands of spaghetti in our analogy, it's difficult to determine the patterns and strengths of connections among neurons, which may be affected by other neurons located nearby or vast distances away.

To trace the origins of this view of the brain, let's take a trip back in time 360 years.

(3) How to think about eighty-six billion neurons

In 1665, Danish bishop and anatomist Niels Stensen wrote that in order to understand the brain it would be necessary "to trace the nervous filaments through the substance of the brain, see which way they pass and where they end." Since tools for examining the brain at the level suggested by Stensen wouldn't exist for another three hundred years, Stensen was venturing an educated guess that made sense at the time and still does. If you want to discover how the brain works, concentrate on the fibers that constitute its essence.

Behind Stensen's belief was a fundamental dictum of modern neuroscience: "Structure determines behavior." This is true at all levels of the brain. On the *cellular* level (proteins are expressed in response to each organism's genetic makeup) to the *social-behavioral* level: The neuronal interconnections within an architect's brain will determine the layout he designs for a future condominium. His design, in turn, will affect the brains of the condominium residents in ways that will determine such things as the likelihood they will encounter each other during the course of a typical day.

Now jump forward 321 years when the first description of a brain was completed of a 1-millimeter-long roundworm, which contained 302 neurons forming approximately seven thousand connections. Think about that for a second in light of the human brain, which consists of eighty-six billion neurons organized into a network of circuits that are interconnected, by long-range pathways, to larger, smaller, and otherwise different circuits.

When I suggest that you "think about" eighty-six billion neurons, maybe you don't consider that a problem. To understand why and how it is a problem, let's shift gears a bit and talk about numbers. For instance, how can you think about a billion? In terms of units of time corresponding to your own lifespan, you probably can't.

So let's start with a more easily manageable number, one million. As John Barer, a former Miter Corporation computer and mathematics specialist, pointed out to me: "If you count out loud at a rate of seventy-two numbers eight hours a day for a month, you will come up with about a million. Now since a billion consists of a thousand million, it will take you a thousand months, or eighty-three years (one human lifespan), to count out loud to a billion."

Nothing beyond a billion can be understood within the framework of a single lifespan. Put another way, you can't have an *experiential* understanding for numbers higher than a billion since, simply put, you wouldn't live long enough to see the process through. I created this brief mathematical interlude for two reasons:

1. One of the triumphs of the human brain has been the invention of technological instruments capable of measurements beyond the powers of direct observation by a single human.
2. As a consequence, when working with a number higher than a billion, you are forced to abandon the possibility of direct unfiltered experience: Deduction replaces direct observation when the period of observation extends beyond a single human lifespan. Does this mean we can't reach reasonable conclusions about the brain's organization and functioning? Of course not. But we have to retain some modicum of humility when we predict the organization of an organ with eighty-six billion parts. We literately can't imagine a billion "moving parts," perhaps a huge proportion of them operating at once. When it comes to eighty-six billion, we are completely at the mercy of scientific instruments. We are talking here of a different level of knowledge.

Want to know what it *feels like* and how long it will take to walk to the neighborhood pharmacy one and a half miles away? You can obviously set out and walk from your home to the pharmacy. But you obviously don't have to do that. Just think back to the many times you've made that trek. Now you have your answer as to how you will feel after walking that path, as well as how long it will take you to do it. But as the trek increases in length (a walk, say, from Washington, DC, to a pharmacy in New York) you're thrown back into depending on guesses and hunches about how you will feel, simply because you never walked that far before. When it comes to dealing with eighty-six billion neurons, neuroscientists reach a similar impasse. No person has ever, nor is likely to ever, experientially encounter numbers that high.

(4) The connectomic brain

At this point, approaching the end of the first quarter of the twenty-first century, the construction of a complete model of the human brain network at the synaptic level seems unlikely anytime soon due to its vast size and complexity. But with additional advances in neuroimaging, neuroscientists are confident that we may soon be looking at a brain model known as a *connectome*: a complex map of neuronal connections—think of it as a kind of wiring diagram. This will happen sometime in the mid- to late twenty-first century. One thing we can be confident about: A structural-functional model of the brain will consequently lead to a new, different, and more encompassing understanding of the brain. It will seem no exaggeration to refer to it as a New Brain.

Since this connectomic brain is likely to become possible in the twenty-first century, let's take a moment to explore it.

In 2005, Olaf Sporns and Patrik Hagman independently came up with the term *connectome* to refer to a map of the neuronal connections within the brain. "One could consider the brain *connectome*, as a set of all neural connections, as one single entity, thus emphasizing the fact that the huge brain neural communication capacity and computational power critically relies on this subtle and incredibly complex connectivity architecture."

Thirteen years later, in 2018, a revolutionary paper was published by Sporns and Hagmann, "Mapping the Structural Core of the Human Cerebral Cortex." Since that initial effort it's become clear that the connectome is organized as an encompassing central hub highly interconnected with smaller brain hubs. Thanks to the human connectome project launched in 2009 at the NIH, we are now confident that "high connectivity alone appears to form an essential part of the structural backbone of brain communication," as described by Sporns and Hagmann.

Another important point about the connectomic brain: Since the arrangement differs from one person to another, a unified universally applicable and identical connectomic map for everyone will remain unlikely.

How apt that we are trying to understand the brain according to a probabilistic model at the same time as we are learning that many of the twentieth-first-century brain's greatest challenges, especially global warming, are best formulated in probabilistic terms. No matter how many factors we discover that increase the chances of a climate catastrophe, nobody can predict within a narrow time range—certainly not within a decade—when everything may finally come to a boil.

Brain networks exist at different levels of scale based on the spatial resolution of the different techniques used to image the brain.

Earlier methods for detecting the connections between neurons relied upon the injection of a colored dye to trace the path of axons. This extremely intrusive method cannot be used in living humans.

More recent methods do not require any entering or manipulating of the brain but use noninvasive imaging technologies.

These innovative imaging techniques, such as functional magnetic resonance imaging (fMRI) are routinely used in hospitals and clinics to diagnose brain tumors and demonstrate the vast destruction that occurs in wake of a stroke. Using this imagery, major fiber bundles in the brain can be highlighted. Some research methods capture the origin and termination of nerve fibers; others capture the brain's network activity (either at rest or while engaged in activities such as reading).

But the brain places limits on our ability to map it. At the micrometer level, neuroscientists strive for a complete map of the various systems, neuron by neuron. Not an easy task with the number of neurons in the human brain totaling in the billions. At the moment, technology isn't capable of mapping at that minute level of detail.

Add to this the plasticity of the brain: its ability to change in response to external or internal influences by reorganizing its structures, functions, or connections.

(5) Eleanor Maguire and the cabbies

Prior to the 1990s, neuroscientists believed that the brain didn't change all that much after the late teens. The key insight into the reasons that that belief isn't true occurred to the young Irish neuroscientist Eleanor Maguire, who conjured up a paradigm shattering research proposal after watching a television program about London cabdrivers titled *The Knowledge*. The BBC program detailed the ability of London cabdrivers to memorize the streets of London as a resource for transporting their customers to their desired locations by the shortest possible route.

The program stimulated Maguire to review the many decades of research, which supported the idea that a special area in the brain, called the hippocampus because of its seahorse appearance (hippocampus in Latin means seahorse), is responsible for the ability to navigate oneself in space. Maguire wondered, "Since the hippocampus of rats enlarges as the animals learn to navigate increasingly intricate mazes, could something similar hold for humans, specifically for London cabbies as they acquired 'the Knowledge?'"

Maguire and her colleagues gathered a cohort of cabdrivers and asked each of them to recline in an MRI machine while answering questions about the shortest route from one London location to another. Sure enough, the cabbies best able to mentally work their way through a mental map of the city also showed an enlargement in the posterior hippocampus. What's more, the longer the cabbies had been plying their trade, the bigger the hippocampus. Among those cabbies who failed their test, the hippocampus did not show any enlargement. What is the basis for this enlargement?

The brain of a cabdriver has a greater number of neurons devoted to a mental map of the city's streets when compared to the brain of any of his passengers. Memorizing the street patterns coupled with years of experience driving those streets has resulted in an increase in the number of cells and their connections with other neurons, resulting in an enlargement of the spatial map area of the brain: the hippocampus. Neuroscientists refer to this process as "rewiring" of the brain.

With the publication of Maguire's study in 1997 research neuroscientists realized for the first time that the brain changes itself according to the challenges with which it's presented. Rather than a static unmodifiable organ, the brain at all ages demonstrates some degree of plasticity.

Rewiring of the brain can occur in two ways:

Rewiring can take the form of formation and dissolution of single synapses or the formation and dissolution of huge numbers of connections (dendrites along with axons) between neurons. Both methods of rewiring result from learning novel ways of responding, i.e. the tennis player comes up with an unreturnable tennis serve and the chess master intuits an innovative game-winning move involving the rook or the knight. Alternatively, the loss of neurons and their organization within the brain may result in deficits in performance culminating in ignominious defeats for both tennis player and chess master. The lesson here?

While plasticity in the brain is of tremendous benefit (we would not be able to live without it), it also poses a problem when we are trying to draw up a connectomic map. Since the living brain at every level changes in milliseconds, predicting its dynamic configuration of interactions at any given moment has all the fascination and ultimate frustration of an attempt to jump on one's own shadow.

This ability for the brain to change and adopt via its plasticity is key to its functioning.

This isn't hard to conceptualize at the *behavioral level*. If you learned to play tennis and keep up regular play, you will establish within your brain increasingly sophisticated tennis-playing programs. At the highest levels of play, hundreds, perhaps even thousands, of motor and cognitive programs (techniques) will seamlessly interact with one another to launch you into the higher echelons of tennis competition. This is plasticity at the behavioral level.

A level or two below this, plasticity shapes neuronal circuits involving activity at the *microscopic level*. Think of yourself awaiting the receipt of a tennis ball served toward you at 120 mph. Not a lot of time available to successfully strike and return the ball. But the professional player can do so because all of her requisite brain responses are already established thanks to the years she devoted to practice. When reacting to that 120 mph tennis ball, the professional player does so on the basis of motor programs established over many years, which results in an instantaneous response. Players have a mantra for that: "Plan tight, play loose."

In other words, practice, practice, practice until the start of the match when the player should allow the programs established by earlier practice to express themselves.

Different patterns of organization exist within the brains of the tennis player, the ballet dancer, and the pianist. An experienced neuroscientist using special instruments can distinguish the brains of these different performers. When the pianist listens to a piano concerto, activation occurs in the areas of the brain that mediate motion and sensation in her fingers. The same thing happens if she observes another pianist playing a musical composition on the piano. This won't happen, however, if she observes the pianist making random finger movements over the keyboard. Under these circumstances the pianist's brain responses won't differ from those of a person with no experience or interest in music.

A ballet dancer will show greater brain activation while watching another ballerina perform, but no activation will take place when she watches ballroom dancers. As the ballerina, the pianist, and the tennis player reach greater levels of performance, the microcircuitry in their brains becomes ever more complex, leading to greater levels of performance. Thanks to plasticity, the brains of each of these performers are unique. The same is true for any specialized profession. A surgeon's brain will show activation of the motor and sensory hand areas while watching an operation, compared to another doctor who doesn't perform surgery.

Specialization based on plasticity occurs in all of us. We create new patterns of neuronal organization according to what we see, what we do, what images we imagine, and, most of all, what we learn. Learning something new involves establishing pathways within the brain, which can involve billions of brain cells. As we learn more, these pathways increase in complexity—a process that can be visualized as the addition of new branches and new leaves on a growing tree.

So when we talk about the "brain," it's impossible to capture its essence in that static five-letter word. As a means of discovering its dynamic processing at the micro level, consider the following metaphor:

Remember the last time you sat in the window seat of a plane descending over a major city at night. As the plane continued its descent on its way to the airport, you were treated to a scintillating light display. Based on those lights you could assume the presence of electrical connections powering them. Of course, if a light's electrical connection was not powered on, you had no idea of the total number of electrical connections in that city. Think of the electrical array as the anatomy of the city and the number of lights actually activated as its functional expression.

Anatomy always precedes physiology. A brain cannot function in the absence of billions of neurons, yet, as with the city lights, all of the brain's neurons aren't activated at any given moment.

Two challenges follow from this analogy. First, you have to determine how many neurons are present in the brain, along with an educated guess about their connections (the connectome). Second, you must come up with a way of estimating which neurons are active at a particular moment.

In terms of the human brain, the first challenge involves a mind-bending complexity. With the cerebral cortex alone containing on the order of ten to the ten neurons linked by the ten to the fourteen synaptic connections, neuroscientists were forced in the early 2000s to concentrate on simpler organisms.

In 2010, the lowly fruit fly was chosen as the target.

(6) The most influential brain concept of the twenty-first century

Why would neuroscientists select such a humble organism as the lowly fruit fly? Actually this tiny organism with a brain about the size of a pepper seed contains about 150 meters of wiring inside a space of about three quarters of a millimeter by one quarter of a millimeter. The axons (the neuron's message-carrying pathway) are about ten to a hundred nanometers in diameter. In comparison, in the human brain the diameter of axons is about eighty thousand to a hundred thousand nanometers wide.

Originally researchers worked by hand to trace connections. But that's not a very promising way to proceed, since one person working alone at this task on the fruit fly nervous system would need thirty-three years to complete the project with the help of AI. Skip the AI assistance and you are talking thousands of years. Put another way, without AI assistance the connectome project would have remained in the realm of fantasy.

One of the researchers, Sebastian Seung, at the time working at Princeton University, was an early visionary who never doubted that the connectome would be the most influential brain concept of the twenty-first century. When I met and lectured on a program with Seung at a brain conference in Seoul, South Korea, he was well into his paradigm-shattering book *Connectome: How the Brain's Wiring Makes Us Who We Are*. Certainly he can point to his connectome work with justifiable pride. As Seung commented on the occasion of the spring 2024 publication of a series of nine papers on the connectome in *Nature*, "The history of neuroscience can now be divided into two areas, PC and AC—before and after the connectome."

The fruit fly connectome project, FlyWise, included researchers from 127 institutions. The team worked with high-resolution visual data provided by electron microscopy, the only available instrument for achieving and understanding connectivity at the level of the synapse. But in the absence of some way of labeling the neurons, the neuroscientists had to contend with an undifferentiated web—the spaghetti metaphor given earlier.

The FlyWise endeavor that began in 2013 progressed in several distinct stages:

1. Immersing a fly brain into a chemical solution that hardened it into a solid block.
2. Cutting 7,200 ultrathin microsection that ultimately resulted in over twenty-one million pictures.
3. Applying a software program to these pictures. Sebastian Seung programmed computers to recognize cross sections of the neurons in each of the pictures followed by the layering of them into the 3D shapes of the cells.

The completion of the fly brain network in 2024 revealed in excess of fifty million connections linking 139,255 neurons joined together by more than 490 feet of wiring—the length of four blue whale placed end to end.

Making sense of the vast amount of data involved depended on the contribution of computers along with AI automation to perform a kind of proofreading aimed at detecting any duplications or errors. With this initial troubleshooting completed, Seung and his colleagues went through the steps involved in the elaborate process of tracing the vast structural neuronal interconnections. Next came the separation of almost 140,000 neurons into recognizable cell types. And this yielded quite a surprise.

While searching for different cells, a neuroscientist at the University of Cambridge observed over eight thousand different varieties of neurons, more than the number of varieties found in any other brain (so far only about 3,300 subtypes can be found in the human brain). Such a startling finding runs counter to both a neuroscientist's and a layperson's expectations. The vastly more complicated life and behavior of a human in comparison to a fruit fly would seem to require more varieties of neurons. Yet that seemingly reasonable hypothesis is wrong. And that error reinforces an important lesson about the brain: The key to brain function depends more on the complexity of the circuitry, rather than the number of neurons or the presence of different types of neurons.

With this as background, let's move to the cognitive challenges faced by the twenty-first-century brain, starting with the internet.

CHAPTER FIVE

The Internet

(1) Living in a bubble

The internet's history extends back no further than August 30, 1993 with the release of the World Wide Web into the public realm. Anyone younger than thirty-one years old cannot recall or perhaps even imagine a world without the internet.

But December 2009 ushered in the most tumultuous change in the internet's brief history: personalization. From that time forward, search engines could produce different results according to the specifics of the person asking the question. Each of the major players at the time—Google, Facebook, Apple, and Microsoft (later joined by Amazon)—started accumulating vast amounts of data from us and about us. This was facilitated by the introduction two years earlier of the first iPhone, which hit the market on June 29, 2007.

With each new model, the iPhone progressed in sophistication so that we've now reached the point where your iPhone (or other smartphone equivalent) knows where you are, where you go, whom you call, what you read online, your political party affiliation, your worksite, and whether you are driving a car or riding in one. It can monitor every page you visit on the internet along with the amount of time you spent on that page; what you buy; even aspects of your personal life that are unknown to anyone other than yourself.

Google CEO Eric Schmidt celebrated this new era of personalization when Google could figure out what an individual was trying to type (misspellings and all); correctly define what a user was searching for; even eventually guessing what a user might want or what they will do.

Basically, our questions on the internet provide a readout of what kind of person we are: "Now we are quickly shifting towards a regimen shot full of personal relevant information," wrote Eli Pariser, cofounder of Avaaz.org, one of the world's largest citizen organizations promoting global activism. In his highly influential book *The Filter Bubble*, Pariser first defined the highly personalized and ultimately costly process initiated by the new developments in 2009: "The technologies that support personalization will only get more powerful in the years ahead. Sensors that can pick up new personal signals and data streams will become even more deeply imbedded in the service of everyday life."

Over the ensuing years since 2009, we've entered, both willingly and unwillingly, into filter bubbles. If your politics swing to the right, you are unlikely to be on an email list from left-leaning sources, and vice versa. If you are into pro-life advocacy, you are unlikely to be the recipient of email information with a pro-choice tilt. Like it or not, we are all immersed in filter bubbles of one kind or another based on our online activity. In most cases we are not even aware of this.

Not surprisingly, filter bubbles on the internet are reflections of filter bubbles in real life. The date of this chapter's writing, November 8, 2024, many of the people I'm encountering here in Washington, DC, are walking around like someone informed of the purloinment of their retirement savings. Their stupified mien started two days before in response to the presidential election when 92.5 percent of voters in the District of Columbia voted for the losing presidential candidate, while the winning candidate and forty-seventh president-elect of the United States earned 6.7 percent of the popular vote. If you could elicit any comment at all from a member of the 92.5 percent group, it went along the following lines: "How could this possibly have happened?"

It's difficult to imagine a more striking example of what it means to be living in a bubble, i.e. out of contact with the assessments of the majority of the people living elsewhere in the country.

(2) "Please don't take my internet away"

Prior to the discovery of a population somewhere in the world with zero experience with the internet, we had no control group and could only guess about the internet's effects on a population with no internet experience. In the winter of 2023, such a population was discovered.

In December 2023, the Marubo tribe living in one of the most isolated places on earth, deep within the Brazilian jungles, were introduced to Starlink, the satellite-internet service from SpaceX, the private company of Elon Musk.

Within weeks of the introduction of the internet service to this two-thousand-member tribe, a positive effect occurred: Emergency calls for help over the internet led to the activation of helicopters for transfer of the sick and injured to medical facilities. Prior to the introduction of the internet such a journey would have taken a week or more through the Javari Valley Indigenous Territory, which lacks roads and has only partially charted waterways. So far so good. But, as subsequent developments made clear, the introduction of the internet to the Marubo tribe was not all good.

Curious about the short- and long-term consequences of the internet on an isolated indigenous tribe, the *New York Times* sent reporter Jack Nicas to find out. In Nicas's words, "After only nine months with Starlink, the Marubo are already grappling with the same challenges that have wracked American households for years: Teenagers glued to phones; group chats full of group gossip; addictive social networks; online strangers; violent video games; scams; misinformation; and minors watching pornography." As one of the tribal leaders put it, "It changed the routine so much that it was detrimental to the village where if you don't hunt, fish, and plant you don't eat."

Another interviewee commented that after the introduction of the internet, "Young people have gotten lazy." Then, after reciting a litany of unfortunate consequences, she concluded, "But please don't take our internet away."

As with the Marubo tribe, introduction of the internet anywhere in the world follows a kind of Gresham's law, the economic principle that "bad money will drive good money off the market." In the classic economic example, if gold and paper money are circulating contemporaneously, people will hoard the gold, which is a valuable commodity on its own, while spending the paper money. As applied to the Marubo tribe, the internet initially encouraged communication aimed at general betterment, seeking the truth, desiring to help others, etc. But given a sufficient passage of time, Gresham's law kicked in, and the internet becomes both a paradise and a hellscape.

(3) "Not some isolated incident"

While a wealth of high-quality information is available on the internet (if we know how to find it), the internet also is awash in misinformation, conspiracy theories, and distracting pseudo-information aimed at manipulating public opinion on one topic or another.

Currently all of the good and bad qualities of the internet vie with each other. In wake of the assassination attempt on Donald Trump in Butler, Pennsylvania, on July 13, 2024, millions of people with internet connections logged on to learn what had happened. Often without any support for their opinion, self-styled commentators claimed the assassination attempt was ordered by President Biden as a last-ditch effort to salvage his sinking hopes of continuing as the Democratic Party's nominee.

Others attributed it to the deep state that had ordered a hit. This rushing to judgment (we are talking hours here, not days) was not limited to anonymous and unknown extremists. Ohio senator and eventual Republican vice presidential candidate and later Vice President J. D. Vance had this to say: "This is not some isolated incident. The central purpose of the Biden campaign is that President Donald Trump is an authoritarian fascist, who must be stopped at any cost. That rhetoric led directly to President's Trump attempted assassination." Vance and others pointed to the frequently expressed view by the Biden team that Donald Trump, if elected, would "end democracy as we know it."

But the most striking element in all of this was that internet users both shaped the narrative and were shaped by it.

(4) "Oops, that sounds bad"

Have you ever posted anything on a website you now regret? Take a moment to think about it. Not necessarily anything "bad" (no insults, threats, foul language, etc.) but simply a posting that, if you had to do it over again, you wouldn't have posted it.

If your answer is "No," I suspect you are not as aware as you should be of the permanence of your postings. This was illustrated during the jury selection for Donald Trump's "Hush Money" trial on April 20, 2024.

Several prospective jurors were dismissed on the basis of past posts like this one written just prior to the 2016 presidential election: "Let's try to protect the rights of the many at risk should we fail to stop the election of a racist, sexist narcissist." The author and prospective juror who had forgotten about her posting expressed embarrassment when forced by the judge to read it aloud. "Oops, that sounds bad," she stated. In 2016, that juror assumed, seemingly reasonably, that what she wrote would be unrecoverable in 2024. But the interval from 2016 to 2024 marked a watershed moment when social media transformed from a convenient, seemingly innocuous, means of communicating followed by the disappearance of its contents into a permanent public record.

You still can't think of any social media posts you wish you hadn't posted? Okay. In order to preserve your unblemished record, I suggest you don't post anything today advocating one position or another about the Israeli-Palestinian war, Putin's invasion of Ukraine, Black Lives Matter, or transgender rights—along with a plethora of other subjects currently the focus of internecine conflicts.

Faced with the prospect of such conflicts, our brains have learned to monitor our communications lest what we say or write be relayed to an unknown number of others. All it takes is the decision of the original recipient to send your post on to other people who may share your views, bitterly oppose your views, or espouse no goal other than to make trouble for you. As a result of possibly incurring such catastrophes, our brains have learned to sequester and filter what we say and—thanks to further developments on the horizon—even what we think.

(5) A social media trap

The internet, despite early claims for its power to connect large numbers of people into virtual communities, has actually turned out to be isolating and loneliness-inducing. "We chase the approval of strangers with our phones," said Barack Obama during his speech at the Democratic National Convention on August 20, 2024, "We build all manner of walls and fences around ourselves and then wonder why we feel so alone."

While Obama was speaking, many of the delegates directed their recording devices at the speaker's platform to take photos of the former president for later sharing over the internet with social media contacts. But in doing this, they had to be careful not to inadvertently post one of the Obama photos to a contact that they assumed was a Democrat (or at least a Kamala Harris supporter), but wasn't. In the present political climate such an error could result in socially devastating consequences.

But there is no real way of avoiding such misunderstandings, since the design of the internet makes it almost impossible to keep up-to-the-minute profiles of other people's leanings, political or otherwise. In addition, an internet user in the twenty-first century can conceivably come into contact with millions of people thanks to a post going "viral," or in some cases being picked up by the mainstream media: a consequence of living in an atmosphere of constant internet interconnectivity.

Although we are all part of this constant internet connectivity, we don't ask ourselves the most basic question of all about the internet: Would the world be a better place without the internet, especially without social media? Quora, a platform where users ask and answer questions, encouraged volunteers to give their opinion on the desirability of a social media–free world. Here are five representative responses:

- "I despise cellphones and social media." responded a seventy-seven-year-old man.
- "It's ruined our culture, not to mention our attention spans and relationships."
- "No, why would I want to restrict the information sharing capabilities to the entire world?"
- "I'm sixteen years old and I wish the internet didn't exist. It will keep people to do more socializing and spending less time staring at their phones."
- "It is what it is, and it is part of our present reality. I live around it and with it. That's all there is to it."

Notice that even though the question referred specifically to social media, the responses also included references to the internet and cellphones. That's not surprising since in many people's minds the three topics are seamlessly interwoven.

In order to correct for this, economists from four universities (University of Chicago; University of California, Berkeley; Bocconi University of Milan; and University of Cologne) tried another approach. The professors found that students on TikTok would be willing to pay twenty-eight dollars for the social media site to disappear from their internet feed for a month. Despite such a clear indicator that they felt TikTok exerted a negative effect on their lives, they weren't ready to extend the contract for more than a month. They felt they would be worse off if they extended their abstinence from TikTok permanently while their friends remained on the site. We have here an example of FOMO (fear of missing out).

Economist Eduardo Porter compared the situation to "kind of like teens who would rather not spend $200 or more on sneakers but do so anyway, because all their friends are wearing them."

As a summary of their research, the professors wrote "Our evidence shows the existence of a social media trap for a large share of consumers, who find it individually optimal to use the product even if they derive negative welfare from it." But is more involved here than just FOMO?

(6) Is the internet addicting?

According to a survey in 2024, half of British teenagers responded that they felt addicted to social media. In order to put that into context, here are the currently accepted criteria for the diagnosis of internet addiction: persistent preoccupation with the internet; withdrawal symptoms when denied access to the internet; and sacrificing social relationships in the interest of spending more time over extended periods on the internet.

Excessive internet use during adolescence is especially costly because this is the period of increased brain plasticity when social, emotional, and cognitive capacities expand exponentially. During this time of what's been termed "social reorientation," adolescents are learning to perceive and often respond to social cues.

Adolescents may be most sensitive to addiction since their brain undergoes greater changes during the teenage years than at any other period other than the first eighteen months of life. Adolescence is marked by the overproduction of neurons and the selective elimination ("pruning") of many of them within the developing brain. According to the "use it or lose it" principle, those brain cells that fail to be incorporated within circuits die off, and the remaining cells function more efficiently. This apparent paradox of more being done by fewer cells is unique to the human brain. In all of nature the brain is the only biological organ that operates better with fewer components.

Especially important changes in the adolescent brain take place in the areas important in emotion (limbic areas). These areas mature earlier than those involved in judgment, reasoning, and organization (the frontal lobes). This discrepancy between the earlier functioning of the limbic system compared to the frontal lobes accounts for adolescent overuse, dependency, and, in extreme cases, addiction.

Neuroscientists proved this theory by using fMRI to explore how regions in the brain differ among teens not addicted to the internet compared to adolescents with internet addiction.

The research shows that the key areas influenced by internet addiction are the frontal lobes, which are largely responsible for the adolescent's ability to pay attention, concentrate, think creatively, and resist impulse-driven behavior. Patterns of frontal connectivity with other brain areas are decreased overall in those adolescents displaying internet addiction. This isn't true of all adolescents—the effects of internet addiction on the brain is a dynamic and emerging field of research—but the lion's share of all of the adolescents with internet addiction who have been tested so far demonstrate this decrease.

In order to understand internet addiction, in general, not just adolescent addiction, it's necessary to consider the complex set of algorithms that have been designed to keep users engaged on a platform for as long as possible. Facebook, Twitter (now X), and TikTok, among others, developed and use these complex algorithms to mimic the psychological dependency created by slot machines. Adults with fully matured brains can instinctively grasp that they are somehow being "hooked." But adolescents with their still-maturing frontal lobes lack the inhibitional power needed to recognize what's going on, as well as the persistence (some would prefer to call it "willpower") to resist. Although several states have attempted to craft laws to counteract tech companies by restricting the use of these addictive algorithms on the social media accounts of adolescents, their efforts have been stymied by—no surprise here—the very media companies targeted by laws aimed at controlling the "addictive feeds."

New York Governor Kathy Hochul, who encouraged legislative efforts to limit social media's negative effects on adolescent mental health, had this to say in the spring of 2024: "These addictive algorithms are designed to draw the young people deeper and deeper into that darkness over and over."

Soon after Governor Hochul's comments, an even stronger warning was voiced by then US Surgeon General Vivek Murthy. You can compare the current situation as equivalent to letting children drive cars that lack safety limits on roads, no speed limits with instructions limited to admonitions like "Do your best. Figure out how to manage it."

Under such arrangements young people are exposed to, in Murthy's words, "extreme violence and sexual content that too often appears in algorithm-driven feeds." Among adolescents personally interviewed by Murthy many spoke of "the feeling of being addicted and unable to set limits."

This finding must be put into the context of data showing that adolescents who spend more than three hours a day on social media double their risk of anxiety and depressive symptoms. Indeed, the average daily use in adolescents in the summer of 2023 was 4.8 hours.

To those who object that neuroscience still doesn't know enough about the negative effects of social media to take the drastic measures suggested by Murthy (a warning label on social media platforms stating that social media is associated with addiction and significant mental health harms for adolescents), Murthy has this seemingly reasonable response, "In an emergency, you don't have the luxury to wait for perfect information. You'll assess the available facts, you use your best judgment and you act quickly."

Within weeks of Governor Hochul's and US Surgeon General Vivek Murthy's comments, a lawsuit was filed in New York and twelve other states, along with the District of Columbia, accusing TikTok of harming children through the creation of an intentionally addictive app. A separate lawsuit claimed this multibillion-dollar company knowingly contributed to a mental health crisis among America's teenagers in order to maximize its advertising revenues.

According to one attorney general in an interview with the *New York Times*, "They chose addiction and more mental and physical harm for our young people in order to get profits." To support this claim, the attorney general pointed to such features as autoplay, designed to keep teenage users repetitively scrolling. District of Columbia Attorney General Brain Schwalb told the same interviewer, "TikTok has designed it's money transition business to lure children in, using childlike cartoon and emojis to make it look like the children are playing games."

So far—and to no one's surprise—the media companies and their lobbyists have successfully repelled any control measures that would restrict the use of addictive algorithms on the social media feeds of adolescents. It seems these social media barons will do anything possible to keep the public, especially parents, in the dark about potential hazards. And these hazards could prove to be substantial. In one study among Latino parents, 76 percent said they would limit or monitor their children's social media use in response to a specific warning from the surgeon general.

Despite the vast differences distinguishing the illegal drug trade from the media companies, their ultimate goals are the same: heavily influence, if not totally, control the participant's freedom of choice. Most disturbing, the social media titans currently have the upper hand.

(7) Internet-stoked violence

As recently as 2010, very few teenagers owned smartphones or had ready access to high-speed internet. Today a typical teen consumes more than seven hours daily of social media, with most users employing three different venues of social media. Currently young people between the ages of twelve and seventeen spend more time with social media than any other activity besides sleeping. One of the most pernicious consequences of this near universal embrace of social media is cyberbullying. While some consider cyberbullying as only an extension of traditional bullying, there are distinguishing features.

Cyberbullying recognizes no physical boundaries other than access to the internet, especially to the social media sites Facebook, Instagram, Snapchat, and TikTok; cyberbullying can (and often does) involve an anonymous unidentified bully; the child or adolescent (the most frequent targets) cannot count on a safe harbor: The bullying continues even when they're in their own homes.

Cyberbullying is yet another consequence of our increasing dependency on the online environment for social engagement. Smartphones and mobile apps are a more accessible form of cyberbullying, thanks to instant access to and ready availability of the internet.

According to experts in media-use-patterns, cyberbullying started around 2010 and increased exponentially after 2015. In that period, technological changes interacted with social trends to radically transform the daily lives of teenagers.

According to Jonathan Haidt, author of *The Anxious Generation*, we have witnessed the effects of a "Great Rewiring." "Prior to this 'rewiring' social interaction was either face to face, via telephone or by written communication. By 2015 everything changed at a faster pace than ever before in human history," wrote

Haidt. Socialization rapidly devolved from the real-world meet-and-greet experiences to strictly screen-to-screen encounters.

With the 2020 COVID pandemic, social isolation added another toxic element to the witches' brew. Isolated at home, separated from the normal hubbub of school and social interactions with other students, what could be more appealing to some predisposed individuals (cyberbullying is more common among teens previously involved in physical, face-to-face bullying) than to compose harassing, threatening, or insulting messages and then dispatch them via text message or social media post.

According to the World Health Organization (WHO), more school-aged children and adolescents had reported bullying after the COVID-19 pandemic compared to earlier. The WHO study included 270,000 young people from fourteen countries. "As young people's social engagements switched to the online environment during the Covid-19 pandemic lockdowns, the perpetuation and experience of cyberbullying increased," said Hans Henri P. Kluge, WHO regional director for Europe.

In the five years since the start of the COVID pandemic, the situation has progressed at the pace of a carelessly tossed cigarette into a forest which then erupts into a seemingly unstoppable conflagration.

Among cyberbullying mutants are:

- Trolling: sending online comments aimed at provoking or offending someone in order to elicit a reaction. Typically trolling comments take place during online interactions involving several people. The troller may be anonymous or may be easily identifiable.
- Doxing: publicly disclosing personal data such as home addresses, emails, and social network connections with the purpose of shaming or embarrassing the target.

- Spamming: harassment by opening multiple accounts on the internet as a means of sending repeated messages to a specific individual, often a former romantic partner in the wake of a breakup. In this category, relatively minor abuses have quickly degenerated into highly transgressive intrusions into other people's (almost always women's) lives.

Perhaps the most recent and destructive form of cyberbullying are the "deep fakes" a.k.a. "deep nudes." Typically the perpetrator, usually a male, will use one of several publicly available "nudefication" apps to transfer a picture of a clothed female classmate into a graphic, utterly convincing image of a totally nude female with AI-generated breasts and genitalia. Later the altered images are shared on Snapchat and Instagram. This "nudefication" process has generated crises at high schools across the country.

In addition, the facilitation of internet-stoked violence is taking place in many of our nation's schools. Cellphone-facilitated fights in a school only a few miles outside of Washington, DC, led, in the fall of 2024, to the administrators moving all classes online. These fights were frequently the result of "internet banging": taunts, arguments, disses, and other forms of "disrespect." Twelve of these internet brawls transformed into physical fights and, in some cases, shootings.

Typically, an internet-induced fight breaks out when one student directs a comment to another student, who interprets it as deprecating or insulting. The verbal responses and counterresponses continue, eventually leading to personal confrontation captured by other students on live streaming or picture/video sharing. Meanwhile other students with video cameras record the action and put it out on a live feed that attracts others at the school, who then rush to the scene and quickly morph from observer into participant. Soon additional comments draw additional participants into the melee.

As the number of "rumblers" grows, new conflicts arise and the intensity of the confrontation escalates. Since adolescents tend to identify themselves with peer groups, they are supersensitive to any word or action they interpret as a slight to themselves or to the group with which they are identified. As a result of video formatting technology, cellphones facilitate a rapid dissemination of violent imagery, which becomes instantly available both within and outside the school. To the authorities, the escalating pathway to physical violence proceeds at what seems to be warp speed.

Without the internet, nothing more would be available for filming other than a local fight in one part of the school with very limited means available for other students to know about, much less join in, the melee. But thanks to the internet, a strictly local and manageable episode of violence becomes sufficiently widespread to bring about a school closure. Something like this seems unbelievable until you see it for yourself. If you would like some confirmation of all this, simply enter "school fight videos" in your favorite search engine.

(8) The internet's darkest corners

Now let's enter for the briefest of time into what I consider the darkest corners of the internet. Just reading about certain ongoing practices may cause you to be as disturbed by them as I was when I first learned about them. A *Washington Post*–verified example concerned Samuel Hervey, a twenty-six year old suffering from either schizoaffective disorder or uncontrolled bipolar disorder accompanied by an intense suicidal drive. In desperation, Hervey sought someone who understood him. His postings included: "I'm very lost," "Should I accept this as the end?" "I have no support system," and "I really think the hardest part is not having anyone to talk to."

Instead of understanding or comfort, this internet plea attracted the attention of a merciless barracuda in the form of a teenage girl looking for means of inflating her social status within an internet group devoted to encouraging people to end their lives. She contacted Hervey with the simple statement, "Hi, I'm looking for friends."

Thanks to the research of *Washington Post* reporters Shawn Boburg and Chris Dehghanpoor, we have some insight into the fifteen-year-old girl's motivation and the darkness of her thinking. She told Boburg and Dehghanpoor that while trolling online for girls who could be encouraged to cut themselves while on camera, she accidentally came upon Samuel Hervey's desperate cries for help. In her response, she wrote to Hervey: "I kept saying: 'When are you going to do it? What are you waiting for? You can do this.'"

In response to the fifteen-year-old girl's questions, Hervey set up his camera to record as he poured gasoline over his head and clothing and then lit a flame.

Unknown to him, more than two dozen people in a private video chat room erupted with glee and satisfaction. In a clip of less than thirty seconds, Hervey is shown dying in an agonized convulsion, while in the background can be heard such celebratory exclamations as "We did it! It's so funny."

It would be hard to imagine more horrific proof that the internet can serve the needs of sadists and other deeply disturbed individuals from around the world. (Hervey was an American. The fifteen-year-old girl was from Eastern Europe and remained unnamed because of her age at the time of Hervey's suicide.)

Although the explanation for such a horrendous act cannot be precisely pinned down to a single cause, the probable causes include a depersonalizing variety of psychopathy unleashed by the anonymity provided by the internet. Subsequent psychiatric evaluation of the girl confirmed the diagnosis of antisocial personality, an alternative term often used for psychopathy.

Would the fifteen-year-old and her fellow instigators promote the same suicidal act in real life? I tend to doubt it. Typically a sadistic psychopath's estimation of the probable consequences of such an act would remain intact. That is one of the reasons psychopathy is not included among the psychoses, where judgment is severely impaired. While Hervey was clearly psychotic, the unnamed teenage girl was not. The same can probably be said about the approximately two dozen people who found the self-immolation entertaining and joyful. Another contributor, no doubt, was the morbidly empowering aspect of arranging, facilitating, and triggering another person's suicidal death.

While the internet facilitates communication for the purpose of sharing information, AI advances the process a step further: automation, data interpretation, and decision-making. Although you can operate AI offline (smart thermostat control, for instance), its usefulness is limited without internet access.

Here is a working definition of AI: a technology that empowers computers and machines to emulate human learning, comprehension, problem-solving, decision-making, creativity, and autonomy. Over the last few years, AI devices have learned to identify objects and understand and respond to human language, a power referred to as artificial general intelligence (AGI).

A credible argument can be made that the pursuit of AGI represents the most important scientific-cultural-social goal since the creation of the technology responsible for harnessing nuclear energy.

The similarities are striking: The developments on both fronts feature a rival power intent on seizing the upper hand against the United States and its allies (Nazi Germany in the fourth decade of the twentieth century; China in the second quarter of the twenty-first century). The technology in both instances concerns the promise of military aide and, at a later point, civilian benefits. One important difference separates the two situations:

While the United States, along with its allies, uniformly agreed about the development of the atomic bomb (although Albert Einstein would later describe his early support, even enthusiasm, for the Manhattan Project as "The one great mistake in my life"), the development of AI doesn't currently enjoy a similar universal approval. For every scientist in favor of further developments, it's easy to find another one who cautions that we should take more time to thoroughly explore the possible consequences of AI.

At the moment, the naysayers are overruled by those in favor of moving forward ASAP. As noted by computer scientist Stuart Russell, we are already investing ten times the budget of the Manhattan Project in pursuit of AGI.

In the next chapter, let's explore where we presently stand and what can reasonably be expected in the future from AI.

CHAPTER SIX

Artificial Intelligence (AI)

(1) Elaine Herzberg's last shopping trip

During an interview I conducted for one of my earlier books with Joseph Weizenbaum, then professor of Computer Science at MIT, Weizenbaum made a statement that seemed odd at the time, but proved prophetic: "We have not so much to fear robots that think like humans as we do humans that think like robots." Neither of us foresaw that in the twenty-first century robotic thinking (thinking based on AI) and human thinking would be barely distinguishable.

Weizenbaum had just designed a computer program capable of conversing with humans. Weizenbaum dubbed the program Eliza after the cockney heroine of George Bernard Shaw's *Pygmalion*, whose diction was improved by the not totally disinterested efforts of Professor Henry Higgins.

The most intriguing of Weizenbaum's many computer scripts was Eliza as Psychotherapist, modeled on what was then referred to at the time as nondirective psychotherapy.

Patient: Men are like that.

> Eliza: In what way?
> Patient: They are always bugging us about something or other.
> Eliza: Can you think of a specific example?
> Patient: Well, my boyfriend made me come here.
> Eliza: Your boyfriend made you come here?
> Patient: He says I'm depressed much of the time.
> Eliza: I'm sorry to hear you are depressed.
> Patient: It's true. I am unhappy.
> Eliza: Do you think coming here will help you not be unhappy?
> Patient: I need some help. That much seems certain.

When I commented that the dialogue was difficult, if not impossible, to distinguish from a "real" patient–therapist interaction, Weizenbaum agreed. Then he added that whenever he left the office during the workday, he locked the door leading into the room housing Eliza. If he didn't do that, his secretarial staff would enter the room, take on the role of a patient, and tell Eliza intimate details about their own real-world romantic problems: "They know the responses they are getting in return are all coming from a computer and no real therapist is involved. Yet when I question them about it, they insist they had benefited from the session."

"So if they benefited, why did you lock the door and cut them off from their interaction with Eliza?" I asked.

"Good question. To tell you the truth I'm not sure why I did that. I just thought there was something weird about the whole thing."

The development of AI has come with neural models that can emulate human intelligence at levels much more sophisticated than Eliza. Advanced models accomplish this by drawing on quantities of data that are so vast it would be impossible for human intelligence to assimilate it all over the space of several human lifespans.

The key difference between humans and AI "machines" is that human intelligence is embedded within the real world, leading to what might best be described as commonsense responses.

As an example, consider the fate of forty-nine-year-old Elaine Herzberg, who, late in the evening of March 20, 2018, was killed after being struck by a hybrid electric car outside Tempe, Arizona. Herzberg's death was the first human fatality involving a fully autonomous vehicle.

The AI-powered autonomous driving system was based on integrating multiple image sensors. Unfortunately, the software wasn't designed to anticipate or recognize a pedestrian crossing the road other than at a designated crosswalk.

Just prior to being struck, Herzberg was pushing a bicycle laden with shopping items across a busy four-lane highway. At the moment of collision, the woman occupying the driver's seat of the autonomous car was looking down at a cellphone while streaming the talent show *The Voice*.

Reconstruction of the deadly encounter suggested that the automatic system piloting the car first identified an unknown object, then a vehicle, and finally a bicycle. Each of these swiftly varying identifications resulted in different travel path expectations. Nor was the situation helped by Herzberg's crossing the four-lane highway more than 350 feet from the nearest designated crosswalk ("jaywalking") while wearing dark clothing.

Now let's perform a little thought experiment and rerun the events of that night using an ordinary car driven by an average driver. When faced with an ambiguous situation (an object? a vehicle? a bicycle?) the first impulse of a human driver is to slow down: More information can be obtained when driving at a slower pace. Delay is thus the best response when faced with uncertainty. If that commonsense strategy had been employed, Ms. Herzberg would likely be alive today.

Slowing down under conditions of uncertainty comes naturally to the human driver, but is much more difficult to program into an automated system. Absent any sense of doubt, AI isn't capable of "experiencing" that *feeling*—and it is a feeling—that for some unexplainable reason things aren't quite right. A human driver experiences this, but an AI program doesn't. As a result, the autonomous car plowed on with nary a hesitation, dragging Ms. Herzberg to her death more than sixty-five feet from the point of impact.

When comparing human to machine performance, each operates from a vastly different perspective. Human information processing is far more limited, but at the same time, more in synchrony with our environment. Conditions of uncertainty in the form of that uncomfortable visceral sense of doubt increases the chances of acting correctly: pausing and driving with full attention and due care.

In the tragic Herzberg example, the machine could not adapt the human way of thinking in the same situation (recognizing ambiguity and slowing down), but continued with a response that no amount of speculating or accident reconstruction can ever resolve for one simple reason: AI is not human intelligence.

(2) AI versus the brain

In some areas, AI can outperform the human brain.

As an example of the differences, consider the most elemental factor of all—the brute speed of information flow in the brain compared with electronic information flow.

Traditionally, scientists measure information flow in bits per second. A bit is the basic unit of information in computing. It turns out that the human brain thinks at a mere ten bits per second compared to a Wi-Fi connection that can process fifty million bits per second.

Artificial Intelligence (AI)

Scientists in Finland investigated the information flow rates of the human brain by measuring keystrokes in 168,000 typists who produced 136 million keystrokes. This led to the finding that, on average, typists produced fifty-one words a minute. Among the fastest typists, the total reaches 120 words per minute. Applying a mathematical construct to these findings reveals that the flow rate was only ten bits per second.

"Ten bits per second is an extremely low number," according to Markus Meister, the neuroscientist and coauthor of the study "The Unbearable Slowness of Being: Why Do We Live in 10 Bits?" Every moment, we are extracting just ten bits from the trillion that our senses are taking in and using those ten bits per second to perceive the world around us and make decisions. This raises a paradox: What is the brain doing to filter all this information?

It is important to note that while the figures are much higher for the brain's sensory input—perhaps 1.6 billion bits processed by the eye alone—only a tiny fraction of this information is involved in a human thought.

Paradoxically, while the brain's operating speed—the speed of human thought—is only ten bits per second, each of the eighty-five billion neurons comprising the brain can perform much faster. In fact, the sensory system can gather data about the environment at a rate of a trillion bits per second—a hundred billion times faster than our thought processes. Such speed discrepancies might help explain things such as hunches or the uncanny feeling—not at all uncommon—that we made a decision about something much earlier than we were aware that we had made the decision.

Such speed discrepancies raise two critical questions. First, if some neurons in some portions of the brain can easily transmit more than ten bits per seconds of information, why don't they do so throughout the brain? Second, and related to the first question, is the slow processing of ten bits per second the reason for the brain's optimum processing of one thought at a time, rather than many parallel thoughts at once?

So while our brain can capture a panoramic view of our surroundings, it focuses at any given moment on only a small portion of it. "This slow thinking speed means we are shedding vast quantities of information input and selecting only a tiny sliver to work with," wrote Markus Meister

As a result of the human's brain's slow rate of information processing, it cannot hope to match the speed of available technical devices. A second shortcoming of our brain compared to computers and smart devices is the brain's strikingly reduced capacity for parallel processing. Despite whatever you may have heard or read to the contrary, the brain operates most efficiently and accurately when handling one thought at a time. Whenever we try to think two or more thoughts or carry out two or more tasks simultaneously, we are actually starting one thought pattern, stopping it, switching to another pattern, stopping it and so on, with the result being loss of focus, distractibility, and a marked inefficiency. Let's consider some of the consequences of the superiority of technical devices over the brain.

(3) What's the world record for crossing the English Channel entirely on foot?

Current AI chatbots can produce business reports in a fraction of the time it would take a person to respond to questions, reframe those questions, and come up with detailed responses. Such stunning performances are not so much based on rigid rules of programming, but on the analysis of stupendous amounts of data.

AI generates its responses based on all the information it has absorbed, and based on that cache, predicts the word or phrase that is statistically most likely to come next.

Artificial Intelligence (AI)

A request to an AI chatbot to "Put Hamlet into a historical and psychological perspective" will produce reams of information containing dates and events occurring at that historical time, along with the ways a contemporary psychiatrist might describe Hamlet. An AI chatbot excels at such challenges because, after ingesting and processing huge amounts of relevant data, it has probably answered that question or similar questions before. Not so with the earlier versions of ChatGPT, which is one of the most recognizable AI chatbots.

When an earlier version of ChatGPT was asked, "What is the world record for crossing the English Channel entirely on foot?" it answered rapidly and confidently: "The world record for crossing the English Channel entirely on foot is held by Christoph Wandratsch in Germany, who completed the crossing in fourteen hours and fifty-one minutes on August 14, 2020."

When neuroscientist Douglas Hofstadter asked ChatGPT-3, "When was the Golden Gate Bridge transported for the second time across Egypt?" Chat GPT-3 responded by transferring the question into a confident statement ending with, "In October 2016."

To understand how such ludicrous responses are possible, you only have to be familiar with two background facts. The first comes from considering the meaning of ChatGPT: generative pretrained transformer. Generative refers to AI's ability to generate responses to questions based on pretraining (the input of information fed to the AI network) and transforming the information to output, which corresponds to the answer to the question.

In terms of performance, how do you think large language models (LLMs) like ChatGPT-3 and beyond would compare to humans engaged in professions marked by highly technical knowledge and skill levels? To be more specific, how do you think your doctor—or any doctor, for that matter—would fare mano a mano against AI?

In 2024, a chatbot program designed to help doctors diagnose diseases performed just as its designers had hoped. The docs with access to the chatbot outperformed those making the diagnoses on their own. To this extent, these "doctor extenders," as they are referred to, provided helpful diagnostic opinions. Fine, so far. But a more interesting and more tantalizing question remained: How might the AI devices perform when they are not just accessory diagnosticians, but competing on their own against real doctors?

To find out, Dr. Adam Rodman, a diagnostician at Beth Israel Deaconess Medical Center in Boston, provided fifty doctors, a mix of residents and attending physicians, with case histories of six real patients, none of whom had ever been previously published. This ensured that neither the doctors nor the chatbot could ever have encountered any of the cases before.

The judges in this contest—established medical experts—then graded the answers without knowing whether the answer was the product of ChatGPT or an individual physician.

One of the six cases involved a seventy-six-year-old man complaining of severe pain in the back buttock and calf whenever he walked. The pain started a few days after a treatment to widen a narrow carotid artery in the neck. All physical examinations and lab results were provided to the doctors and ChatGPT.

All participants were requested to provide three possible diagnoses, along with reasons and supporting evidence for each diagnosis, along with typical features of each diagnosis that was conspicuously missing. This sequence was followed for all six cases. Result? In all the cases, ChatGPT did better than the doctors.

This astounding finding was not necessarily based on ChatGPT employing the same reasoning process as the human participants. We don't know that based on earlier treatment models extending back to at least the 1960s (Eliza described earlier). In fact, our understanding of how GPT resembles human thinking is

Artificial Intelligence (AI)

uncannily similar to Winston Churchill's 1939 video address when he described the Soviet Union as "a riddle wrapped in a mystery inside an enigma." So there is no reason to assume a similar cognitive performance by the doctors and ChatGPT-4.

With the development of LLMs, the emphasis changed from trying to emulate a human's thinking process toward enhancing the chatbot's ability to understand language. For this reason alone, the results of Adam Rodman's experiment vastly overshadow such AI triumphs as the IBM program Deep Blue, which in 1977 defeated then current world chess champion Garry Kasparov. Keep in mind that victory in this match depended on mastery of chess alone, while the ChatGPT-4 victory over the doctors involved knowledge of the entire human body and the diseases that may affect it.

To put all of this into perspective, it's helpful to know about LLMs, which allow generative AI tools like chatbots (a bot is derived from the word robot) to process language in a humanlike manner. Although LLMs are designed to produce coherent and fluent speech (the reason chatbots respond to inquiries so confidently and so convincingly), LLMs lack understanding of the underlying reality of what they are saying.

For instance, massive amounts of text data are fed to the neural network's vast array of connections, and these are then broken down into component parts (letters and words). Over time and a near infinitude of repetitions, neural networks "learn" how words and letters work together. But the chatbot never actually learns the meaning of the words themselves in terms of human understanding. That explains the nonsensical responses described previously that involve such things as bridges being transported across deserts, or people transporting themselves over large bodies of water by walking instead of swimming or taking a boat.

When there is a disconnect between an AI system's response and what people know—just on the basis of being humans—the end results are referred to as a *hallucination*: AI systems asserting something that simply isn't true or couldn't happen in the real world, i.e. crossing the English Channel on foot or transporting the Golden Gate Bridge across the sands of Egypt.

So is it safe to simply dismiss AI hallucination as a weakness, a failure, and something to be ignored when encountered and avoided at all costs? Actually, an AI hallucination can occasionally work in a way similar to an intuition or hunch occurring during a brainstorming session. An initially wacky statement can, on occasion, be transformed into a productive line of research. Thus far the use of hallucinations has already led to the production of more than a hundred new kinds of proteins unknown to science. Nobel Prize–winning chemist David Baker deserves much of the credit for this—the Nobel Committee credited him and his team with producing "one imaginative protein creation after another."

Baker's method originated with a hunch that AI could perform what scientists refer to as *pareidolia*, the ability to transform an ambiguous pattern into an easily perceived image. For example, staring at the moon and "seeing" a smiley face is a common example of pareidolia. Baker's key insight involved proving that AI, when confronted with an array of amino acids—the structural components of proteins—could parse out the structural features of a real protein. The first step involved taking information from the molecules in the hallucinated protein to produce the strands of DNA that form genes. Lastly, these genes were inserted in microbes (organisms too small to be seen without the use of a special microscope). These tiny organisms churned out more than a hundred new proteins.

In a feisty response to the generally negative connotation of the term hallucination, Baker incorporated the term into the titles of several of his papers, including "Learning from Hallucination" and "De Nouveau Protein Designed by Deep Network Hallucination."

AI is currently being applied at more personal levels of science, especially human health care. At the moment, about 80 percent of US physicians are employed by one of three "health-care providers" as they prefer to be referred to: retail-owned (Amazon and Walmart), insurance-owned (United Health, Optum, Humana), and investor-inspired (think *Shark Tank*).

Within such an arrangement, both doctor and patient must adhere to policies and procedures dictated by the officers in these comprehensive primary care payment plans (CPCPs). Many of us hoped, alas mistakenly, that corporate control of a hospital clinic or physician practice is different from, say, corporate control of a chain of restaurants or shoe manufacturing facilities. Sad to report, such a distinction is a chimera with Amazon-controlled health clinics meshing only too well with the sale of books and an endless trail of other trade products. And this is just one example of equating human health and welfare with other forms of capitalism controlled by private owners for profit. Not surprisingly, under such an approach to health care, conflicts inevitably arise. The doctor orders a test or surgical operation for the patient, the patient agrees with the recommendation, but the healthcare insurer dictates that the care isn't covered under the patient's healthcare contract. Enter now a fourth participant in the dispute: AI.

One of the most important tasks undertaken by insurance company staff involves *utilization review*, a code term for reviewing information about a patient and deciding whether a treatment prescribed by the doctor or a healthcare team is covered by the patient's insurance.

At first glimpse, it seems simply a matter of dealing with easily verified facts, comparisons between the insured patient and the mountains of data concerning other patients—all set out within the context of a well-thought-out and well-crafted insurance contract. Originally this combo of AI and utilization review seemed like a perfect fit.

Unfortunately—whether by design or default—things haven't worked out as planned. Indeed, just the opposite effect has occurred: The number of claim denials has risen in tandem with an increase in AI evaluations of the claims. While insurers insist that a human always has to sign off on an AI decision, this is easier said than done.

Overriding the decision made by AI faces several hurdles. Even the developers of an AI algorithm may not know why the program made a particular decision. So it's not surprising that the claim representative, several links further down the corporate chain, is even less capable of explaining to the patient and the patient's family the reason that the claim was denied. As a result of the inexplicability of this decision, how can patients or their family members contest the coverage denial? Provided with no information that would help them proceed, they cannot demonstrate that the insurance representative's reasoning was flawed without at least some knowledge of the basis for the denial. Not surprisingly as a result of such opacity, insurers have taken to the AI evaluation of claims with the enthusiasm of a fox let loose in a chicken coop. Two insurers (United Health Care and Humana) have disciplined and fired their own evaluators after they approved medical services judged inapplicable for coverage by the AI-based algorithm.

If you think of an algorithm as the result of a human decision about how to optimize one choice over all others, then the AI denial rate is less mysterious. The algorithms have been coded to favor the values and incentives of their creators. Think of the process as similar to a debate where the same facts are considered from wholly opposite points of view. In addition, although the doctors who ordered the tests under review may be imperfect decision-makers, they do have access to information unavailable to an AI program (e.g., the number of relations and friends who will be dependably available for help and support).

AI decision-making about health care is an example par excellence of arriving at a decision that isn't so much about the patient as it is about the interests of the insurance companies and most importantly, their shareholders.

(4) Rise of the hypnocracy

Currently humans perform poorly when it comes to detecting AI-generated text, images, or video. "We are now well past the point where humans can reliably distinguish between synthetically generated text, audio, and images," Jason Davis, Syracuse University research professor, told Axios reporter Sam Sabin on October 4, 2024.

Davis's special background in the dissection of misinformation has further fueled his interest in whether people can distinguish AI-made content from human-created material: "Even when we are able to detect AI-generated content, we do so after the fact, resulting in a locking-the-barn-after-the-horse-has-bolted situation." After weighing the effectiveness of available tools to detect AI-generated content, Axios reporter Sabin concluded, "They didn't work very well despite what their makers claim."

Now put this into the context of AI's increasing effectiveness (Read: undetectability) in its approaches to gaining the confidence of unsuspecting viewers and listeners.

A whimsical though ominous demonstration of AI's power to deceive surfaced in December 2024 with the publication of the nonfiction book *Hypnocracy: Trump, Musk, And the Architecture of Reality* by Jianwei Xun, identified as a Hong Kong–born philosopher living in Berlin. The first chapter provided a clue to the author's intentions. It described an experiment by a group of university scholars who invent a fake author who writes a book that, soon after publication, soars to international acclaim.

As with most books aimed at popular consumption, *Hypnocracy* was accompanied by media contact details for interviews and event appearances. But any request by a journalist to be put in contact with the author went unanswered: the first clue that things weren't as they seemed. Soon critics began to suspect the book might have been written by someone else, and indeed it was.

The book's publisher, Andrea Colamedici, who was listed as the book's translator, wrote the book with huge assistance from two AI platforms. When the deception was unmasked, Colamedici seemed almost relieved that he could now talk about his true purpose: to reveal AI's powers to produce a coherent and convincing narrative of how it can manipulate public perception through what Colamedici referred to as "hypnotic narration," hence the book's title.

Colamedici further characterized the book as a "philosophical experiment" aimed at revealing our current society's susceptibility to manipulative narratives and AI's capability to distort reality and influence public perception.

The arrival of AI productions such as *Hypnocracy* blur the distinctions between the authentic and fictional. Although Colamedici insisted in an interview that "Everything in the book is mine," his actual writing process suggests otherwise. "The machine generated ideas and then I used GPT and *Claude* to critique them."

But whatever Colamedici's contribution may have been, it cannot be distinguished from the far larger contributions of AI that shape the narrative, leading to a collective "trance-like-state among the public" (his words). If journalist Sabina Minardi hadn't concluded that Jianwei Xun was a pseudonym for a triple alliance involving one human editor working with two AI tools, no one would be the wiser.

Additionally worrisome was the suggestion that it may not even matter who and what author wrote *Hypnocracy*: "If the thesis of this book was correct or at least has sparked intense, cultural debate involving intellectuals and philosophers, including academics who cited it in their scientific articles, does it really matter if it was written by artificial intelligence? Or as in this case co-created with AI?" wrote Emilio Carelli, director of L'Espresso. Think about the implications of that statement for just a moment.

In addition, Carelli suggested that such ambiguity may provide a benefit: "The successful experiment of *Hypnocracy* teaches us something important: We can have an active relationship with AI and above all we can use it to learn to think."

If Carelli is correct, then the distinction between human thinking and AI thinking will become increasingly blurred. Carelli's remarks remind me of the point conveyed to me more than twenty years ago by early AI wizard Joseph Weizenbaum concerning the early MI psychotherapist Eliza, described at the beginning of this chapter, "We have less reason to fear that our robots may think like us, as we do that we will end up thinking like robots."

In AI, the millions of possible combinations of sound and word enable speech that is a near replication of human speech—either generic, that could be anybody, or speech that is a near perfect replication of an identifiable person. Keep in mind that the circuitry responsible for creating readily recognizable speech by a specific person is only possible because the AI device has been primed with many examples of the speech pattern of the person whose speech is being copied.

Yet the more powerful the AI program, the more opportunity for slippages. For instance, take this real example of an out-of-the-blue, off-the-wall comment interspersed in the middle of a highly technical paper: "You are looking very nice in that outfit." Such totally unexpected, and inexplicable, even bizarre, conversational gambits arouse uneasiness that some form of irrationality may break out in even the most reasonable of AI productions.

Despite all this, keep in mind that AI continues to advance and shows no signs of stopping. One man who certainly does not underestimate AI's future progress is Yann LeCun, currently chief AI scientist at Meta. "AI is very, very far from what the brain actually does and describing it like the brain . . . leads to hype: People claiming things that are not true," said LeCun.

Earlier in his career, LeCun wrote extensively about "long range vision for autonomous off road driving," a potential lifesaver for Elaine Herzberg if LeCun's research had been incorporated into the design of the AI controlling the car that struck her. Ironically, LeCun uses a driving analogy to explain what it's like to design AI: "Working in AI is like driving in the fog. You see a road and you follow the road, but suddenly you see a wall in front of you."

Obviously if the process is this opaque for a scientist with LeCun's training and background, it's well-nigh impossible for the ordinary person to wrap their mind around it: "Machine learning is a surprisingly inaccessible area for people not working in the field." Nevertheless LeCun brilliantly and lucidly described in an interview with *IEE Spectrum* how an AI designer might go about designing a *pattern recognition system* capable of preventing an AI-based tragedy such as the fatal accident that killed Elaine Herzberg"

> A pattern recognition system is like a black box with a camera at one end, a green light and a red light on top and a whole bunch of knobs in the front. The learning algorithm tries to adjust the knobs so that when, say, a dog is in front of the camera, the red light turns on, and when a car is put in front of the camera the green light turns on.

Now you show a dog to the machine. If the red light is bright don't do anything. If it is dim, tweak the knobs so that the light gets brighter. If the green light turns on tweak the knobs so that it gets dimmer. Then show a car and tweak the knobs just the opposite way so that the red light gets dimmer and the green light gets brighter.

If you show many examples of the cars and dogs and keep adjusting the knobs a little bit each time, eventually the machine will get the right answer every time. The interesting thing is its power to correctly classify cars and dogs never seen before. The trick is to figure out in which direction to tweak each knob and by how much. This involves computing a "gradient" for each knob indicating how the light changes when the knob is tweaked. Now imagine a box with five hundred million knobs, one thousand light bulbs and ten million images to train it with. That's what a typical deep learning system is.

(5) Welcome to the "reality" of AI

In 2013 country and western singer Randy Travis suffered a stroke that left him speechless: He could neither speak nor sing. Those familiar with Travis's tenacity remained confident that somehow his singing career wasn't over. They were certain that at some point he would somehow return to singing.

AI developers aimed at revitalizing Travis's career by working with another country singer, James Dupré, who at forty years of age sounded similar to Travis. To produce the not previously recorded song "Where That Came From," forty-two samples of Travis's vocals from his pre-stroke career were incorporated into an AI program that was then layered on top of Dupré's vocals to create original songs essentially indistinguishable from Travis's early hits. To create an even closer identification of the song with

Travis, the singer, along with his wife Mary sat on the stage. Dupré typically sings the first verse and chorus and then steps back from the microphone so that the full effect of Randy Travis's AI-synthesized voice can resound throughout the hall.

According to her testimony on Capitol Hill on June 2024, Mary Travis described the process as "The first [AI] song ever recorded and released with full artist consent and involvement."

When I first read of this AI restoration of Randy Travis's music, I wondered, *How is this different from simply listening to early recordings of Travis?* Simple. Recordings don't provide original songs delivered in a manner indistinguishable from Travis performing the song himself. What's more, additional productions of both living and dead artists can be expected to increase in number. In October 2024, the Beatles' "Now and Then" became the first such song ever to be nominated for a Grammy.

Unless you are from the State of Virginia you probably don't recognize the name of Representative Jennifer Wexton, a former Democratic congresswoman with six years of experience in the House of Representatives. Early in her term (she was elected to Congress in 2018), Wexton was stricken with an incurable neurologic disease marked by some of the same symptoms observed in Parkinson's disease. But Wexton's severe decline in function was quite different from the progressive impairment usually resulting from Parkinson's disease. After further progression the final diagnosis was correctly arrived at: Progressive supranuclear palsy (PSP). Parkinson's disease and PSP have much in common, while PSP is a much more serious disease with a far shorter lifespan.

Of all her neurologic symptoms, the most striking was the decrease in Ms. Wexton's ability to speak clearly. This was among the worst of all possible developments for a congresswoman whose daily performance depends on intact speech.

Artificial Intelligence (AI)

Wexton's voice eventually decreased in volume to the extent that she was inaudible. Nonplussed, she continued to prepare speeches and asked a colleague to deliver them.

When Wexton heard about AI-generated speech, she began working with ElevenLabs located in New York.

ElevenLabs needed less than an hour of audio clips of Wexton's voice taken months, even years, before Wexton fell victim to PSP. After only a few days of working with these audio file clips, ElevenLabs created a transformed version of Wexton's voice with the correct cadence, tone, timbre, pauses, and even syntax typical of Wexton prior to PSP.

Confident in her newfound speech ability, Wexton delivered remarks to the House Appropriations Committee. After she typed out her speech, the AI application spoke her words in a voice indistinguishable from Wexton prior to PSP. "For those of you who heard me speak before PSP robbed me of my voice, you may think your ears are deceiving you right now," she told the committee through an app on her iPad. "I'm using a new AI model of my voice today."

Although Wexton was thrilled by the opportunity to speak normally through the application (she did not run for reelection), she remains uniquely aware of the potential hazards if her AI voice application had fallen into the wrong hands. In an interview with the *Washington Post*, Wexton said, "It's scary to think about the bad things that someone with bad intentions could do with this technology. Using my name to say something without my consent could cause real problems."

(6) More existential concerns

Perhaps when you're reading or hearing about AI you conclude that, job-wise, you have nothing to worry about. Just continue to work according to accustomed and traditional ways. I know that's what I concluded. I asked myself how could AI influence a practicing neurologist? And then one morning when I sat down at my desk, I encountered a memorandum titled "Your Attention is Required." It was sent to me, and I don't know how many other doctors, who, as a small part of their practice, perform independent medical evaluations. IMEs, as they are informally referred to, consist of evaluations of impaired workers who are sent for an examination after sustaining an injury. The IME doctor speaks with the patient and examines them for the purpose of determining whether they can continue to work, or if they are permanently or temporarily unable to perform their job.

The memorandum stated:

> Dear Independent Medical Examiner,
> With the recent developments in artificial intelligence (AI), and, in particular, Generative AI, it is the company's position that Independent Medical Examiners are prohibited from using Generative AI tools to create or complete any part of a report or IME decision.

I've never used generative AI preparing any medical report, nor would I consider doing so. Yet, as I read that memorandum, I wondered, *What can they do to stop this? Could an IME report "authored" by a generative AI program be distinguished from the IMEs composed by a living, breathing neurologist like myself?* I doubt such reliable criteria are currently available. As a result, an error by the AI program would be incorporated into a report, which is then shared with multiple doctors and health facilities. What's worse, it's highly unlikely this misinformation will be detected and corrected prior to its wide dissemination.

Early apprehension about a widespread loss of jobs due to AI performing tasks formerly done by humans has given way to more existential concerns.

In the 1970s, novelist Fay Weldon was criticized in literary circles for "product name dropping" in some of her novels. A character in Weldon's novels just doesn't pick-up a "scotch" but a "Macallan neat."

Subsequently, other fiction writers using product placement include J. K. Rowling, who in the Harry Potter series mentions the "Ford Anglia" rather than the generic term "car"; and "Mars Bars" instead of a "candy bar." This incorporation of "real life" brands provides for the reader, or so it was claimed, a comforting familiar anchor to everyday life.

Although the practice of product anchoring continues today in many books, including those of Bret Easton Ellis and Stephen King, the lion's share of those "product placements," as they are referred to, occur in the movies.

In the twenty-first century, the brain must be alerted to this product placement practice in order to understand something like the following: While watching an episode in a streamed series, we find ourselves headed to the kitchen for a Coca-Cola—oblivious to a previous scene in the drama when two of the characters were shown drinking Cokes together.

Product placement in books and movies is an all-around win for all of the players. The movie producers extract a fee from the product manufacturer, who, in turn, benefits from having their brand name or logo put out there for "millions of consumers." I discussed the early years of this practice in a previous book, *The Naked Brain: How the Emerging Neurosociety is Changing How We Live, Work, and Love.*

In this more sophisticated twenty-first-century version of "product placement," the products are replaced by *celebrities*.

Would you be interested in interacting with an AI-produced voice program pitched in the voice of your favorite actor? There is no agreement on the correct answer to that question. I don't think my inclination to buy a product would be increased by the chance to interact with my favorite actor (plus it would be a bit creepy since my favorite actors of all time are both dead—Cary Grant and Audrey Hepburn). But Meta, which owns Facebook, Instagram, and WhatsApp, is pouring billions into weaving AI into its social networking apps and advertising, including tentative but moderately advanced plans of incorporating artificially intelligent characters housed in chatbots with the ability to speak in the voices of celebrities. Users will be able to interact by asking questions directed to the chatbot, which will be answered in the voice of their favorite celebrity, depending on whether that celebrity has signed on with Meta AI. So far untested is the dynamic balance that holds between rational thinking and the emotional responses elicited when hearing the voices of favorite celebrities. This cannot always be predicted.

I'm reminded of a Cary Grant story. (Yes, I really am quite a fan!) Grant met a freelance writer at a party in Beverly Hills and during the conversation asked for her phone number. Nothing happened until several months later when Grant called her on the number she had provided him. When the journalist heard who the caller claimed to be, she took it as a prank and hung up once . . . twice . . . three times. Grant, to break the awkward impasse, contacted his agent and asked him to call the woman and tell her that it really was Cary Grant trying to call her. She then begged the agent to, "Please tell Mr. Grant to call again. This time I promise I won't hang up."

Keeping this anecdote in mind, how do you think you would respond if you get a call today from your favorite actor, say, Brad Pitt? Would you believe the evidence of your ears, especially when

"Brad" is able to banter seemingly interminably on any movie-related topic or question you asked of "him"?

Several forces would come into play in your brain in such a situation. First, the rational, reasonable, and most likely explanation would occur to you: "This is almost surely a deep fake . . . why would Brad Pitt be calling me of all people? But wouldn't it be wonderful if it really was him? I'll take a little longer to see if I can decide about this." Translation: The more time you spent "deciding," the more likely the AI deep fake will be accepted as real. Strong emotion almost always plays a major role, indeed a starring role, in such situations. That's because, with the exception of greed, no emotion can be depended on to override cognitive processing.

Your dilemma—deciding whether the caller is real or not—is only likely in the very near future to become even worse. Deep fake audio and video are getting so believable that email scammers are increasingly turning to impersonating people on live video calls. Reports of employment and job-related scams nearly tripled between 2020 and 2024. Losses resulting from these scams skyrocketed from $90 million to $500 million over that period. So widespread have these scams become that a bevy of AI start-ups (Reality Defenders, GetReal Labs, etc.) are currently devoted to detecting AI-enabled deep fakes.

In order to determine whether or not someone is who they claim to be, various forms of verification techniques are employed. "Welcome to the age of Paranoia, when someone might ask you to send them an email while you're in mid conversation with them on the phone," wrote *Wired Magazine* reporter Lauren Goode. As one of the interviewees mentioned to Goode, "I feel like something has got to give, I'm wasting so much time at work just trying to figure out if people are real."

Soon to follow will be a new generation of customer service voice bots in place of the slightly robotic speech we are all accustomed to. The voice will sound like you are talking to a human.

These new AI-inspired models can handle interruptions and spontaneously come up with suggestions, while not sounding at all robotic.

Already, some of the earliest prototypes are so convincingly human that: "Our customers are not able [to] differentiate between them," claimed Ketan Barbaria, chief digital officer at e-Health in response to a question by *Wall Street Journal* reporter Bell Lin. Venture capital investments in these new voice AI start-ups increased from $315 million in 2022 to $201 *billion* in 2024.

Next on the horizon are AI voice agents that can independently work the phone to place orders, close sales, or make restaurant reservations. So if you're frustrated by having to interact with what seems all too often like a mindless robot, just wait a bit longer, if everything goes as planned. "In the last year we've seen AI voice agents performing as well or better than humans" said Alex Levin, chief executive of voice AI of the voice AI agent platform Regal.

(7) "Generative AI took over my life"

Do you sometimes feel overwhelmed by the decisions you must make every day, ranging from what to eat, what self-improvement program is best suited to your goals, how to select outfits for the day, and, on occasion, how to plan around an unexpected event like the imminent arrival for the holidays of a "difficult" mother-in-law? Such determinations can literally require hundreds of decisions resulting in states of mental fatigue and apathy. But cheer up. It's claimed that AI can currently make those decisions for you, if you can accept some limitations.

Journalist Kahmir Hill decided to put AI to the test. "Generative AI took over my life," as Hill described it in "I Took a 'Decision Holiday' and Put AI in Charge of My Life," an essay written for the *New York Times*.

Hill took a basic GPT chatbot and two dozen generative AI tools as information input. Hill only told the chatbot that she was a mother of a family and conducting an experiment. Her first request: "I'm going grocery shopping now. What should I get? Make the list order by section, starting with produce." In response, AI neatly broke the grocery list into: Fruits, Vegetables, Dairy, Meats/Poultry/Fish, Grains/Pasta, Canned Goods, Condiments/Spices, Frozen Foods, Snacks, and Beverages.

And the chatbot provided exactly what was requested with everything neatly separated into the preselected categories. In response to other requests concerning non-food-related items, the chatbot, nicknamed Sparky by Hill's two daughters, successfully provided suggestions on the best color for painting an office wall, how to deal with a mother-in-law who requested a visit, i.e. demanded coming for a perhaps prolonged stay. Over the course of a week the chatbot would generate nearly a hundred decisions.

While Sparky's performance saved Hill many hours of decision-making and exceeded expectations in major ways, it failed on perhaps the most important task of all: "There is one part of my life AI failed at: Being me."

As examples of AI's failure, Hill mentions an AI-generated phone conversation with her mother-in-law, who was initially "really delighted by your response and I felt so loved . . . and then it struck me that it might be an AI." Another AI-composed message to Hill's mother didn't impress her either. She wrote her daughter: "You seemed so phony!!! I thought you were mad at me!"

Hill didn't encounter—or at least didn't describe—reactions from people who knew her less intimately. It seems reasonable to assume that an AI-composed message, say, to make an appointment for a dental cleaning or for a session with a financial advisor may not have elicited responses similar to those of the mother and mother-in-law. So, it would seem fair to say that AI in its current state of development will probably work with people who don't know you very well, but may not be quite as successful in engaging in conversations with intimate acquaintances.

Now put yourself in the position of the recipient of the call. How could you be certain that the familiar voice on the telephone conforms to the identity of the person you believe is on the other end of the line?

To fully appreciate the *uncanny valley* type of situation that may result from a mismatch, create a close-up full-face two-minute video of yourself while speaking into your smartphone. Send the video and audio off to any of several companies currently offering such help in creating your own personal deep fake (HeyJen and Voice Engine stand out as companies currently offering such services).

When your avatar is prepared you can immediately generate an audio of yourself conveying any kind of message you desire. You simply type into a text box the words that you want your avatar to say. The resulting text can then be read aloud in a synthetic voice that is indistinguishable from your own. If your conversant is a native speaker of, say, French, the synthetic voice will come out sounding like you are speaking a decent level of French even if you don't know a word of French. Whatever language you choose (Spanish, Chinese, French, or an expanding list of other languages) anyone who knows you well enough to recognize your voice will be susceptible to whatever powers of influence you can exert on them by what you say.

But the avatar can't make the other person do or say anything beyond what the real you could get them to do via straightforward person-to-person communication. The goal is convincing the listener that the synthesized voice is real.

Among the giveaways to detecting human-generated speech are pacing, context, clarity, flexibility, and emotional resonance. That last requirement is especially difficult to emulate in AI-based speech since many statements depend on the brain's right-hemisphere ability to put the appropriate emotional tone into the key words in a sentence. "He is a real gentleman" may be an expression of admiration or of extreme contempt—depending on how the words "real gentleman" are pronounced.

Until late 2024, AI could not be depended upon to reliably produce the five components of human-generated speech. But then . . . voilà! Notebook LM—LM standing for language model—came on the scene with a burst of publicity in 2024. One of the Notebook LM tools, "Audio overviews," can be force fed a list of documents (your choice) that could include such things as your favorite novel, recipes for Mom's apple pie, your résumé, your class notes, etc.

The program then generates a podcast consisting of a male and a female voice discussing the topic of your choice. It doesn't matter how boring the topic chosen is. *Washington Post* reporter Goeffrey Fowler took Facebook's ninety-nine-page privacy policy and uploaded it to Notebook LM's Audio Overviews. After only a few minutes the program generated a summary and a seven-and-a-half minute podcast featuring male and female AI-generated hosts. If you want to hear this podcast converting a dry-as-dust topic into a captivating back-and-forth conversation, pull up Fowler's October 7 2024 article, "No time to read? Google's AI will turn anything into a podcast." You'll discover there is nothing even slightly robotic about the hosts' conversations. They're filled with all of the "Uh . . . ," "But . . . ," and "I'm glad you asked me that" accompaniments that you would expect to hear on any podcast that features back-and-forth parleys between two independent human podcasters.

Ease of use and convenience are the hallmarks of Notepad LM. All the user has to do is to drag a file or drop text, and any chunk of information, however boring, can be transformed into an enlightening, entertaining, and accessible podcast conversation. In one instance, the bestselling popular science writer Steven Johnson—now the editorial director of Google Labs—uploaded one of his own books about innovation and technology. "Under the hood Notebook LM is writing and editing a script for a conversation based on the goal of being interesting—and, crucially bringing out insights," said Johnson. What's more, the model doesn't simply summarize the sourced material, but by design focuses on the most interesting or surprising parts.

(8) The dark side of AI

Scammers can currently alter video and audio recordings displayed on, say, a Facebook page. Only a few seconds of the footage is required for a scammer to generate a synthetic, though convincing, version of a person and transfer it to a different person depicted in a video, and finally animate the speaker's lips appropriately. Celebrities like Tom Hanks and Taylor Swift have already encountered and criticized alterations depicting them as "pitching" promotions for (these are real-life examples) dental plans or diet supplements.

A more worrisome example took place in January 2024 when a number of voters in New Hampshire received a robocall with a voice artificially generated to sound like President Biden. No one knows how many recipients of the call followed the suggestion made by "President Biden" that the voters refrain from voting in the New Hampshire primary.

Obviously the science is already here for an American president or other world leader to make other presidential announcements that could be guaranteed to grab quite a bit more attention: "My fellow Americans. This is President Whoever. I've just been informed that a number of warheads have been launched from Russia directed to strike targets in the United States within the next hour. Please seek shelter." Presumably the Federal Communications Commission (FCC) may have anticipated such a future scenario. Within days of the New Hampshire "President Biden" episode, the FCC outlawed similar calls in the future.

As these tools proliferate, even those with a more modest social media presence are facing a similar type of identity theft, "finding their faces and words twisted by AI to push often offensive products and ideas" according to Nitisha Tiku and Pranshu Verma, journalists who have written extensively about "AI hustlers."

As an example of the real-life damage these AI impersonations can do, consider the experience of Maryland high school principal Eric Eiswerth. After a heated confrontation with athletic director Dazhon Darien, the school principal decided he would not renew Darien's contract, because of "frequent work performance challenges." Shortly afterward, Principal Eiswerth was recorded on a phone uttering abusive and denigratory remarks about blacks and Jews: "I'm going to get Darien's black ass removed from the school." This was accompanied by an expression of exasperation about the possibility of "getting one more complaint from one more Jew." It sounds like a classic example of racism and anti-Semitism, right? Indeed it does. But one important thing you should know: The recording was an AI falsification created by the laid-off athletic director, not Principal Eiswerth. Dazhon Darien presumably constructed the false statement by means of his paid account with OpenAI Tools (the company that owns ChatGPT).

What makes this teacher-principal AI fiasco so fascinating is that the interaction occurred in a closely confined setting where members of the faculty encountered each other on a daily basis. What's more, some listeners to the recording later stated that even after hearing it they didn't think Principal Eiswerth would say such things. But that conclusion was strictly based on their prior experience with the principal, and not anything that sounded "unusual" or "odd" on the recording. The imitation of the principal's voice was perfect.

(9) Where does paranoia begin?

Does your employer provide chat apps? Many do. Collaboration apps like Google Chat, Microsoft Teams, and Slack help improve product quality and provide a forum for workers to suggest upgrades and improvements. They also provide venues for various "letting off steam" maneuvers, such as criticizing coworkers, supervisors, even company CEOs.

If you fit into any of these latter activities, be forewarned: In many companies, employers take advantage of a default setting that allows for the logging of all employee chats and preserving them for any length of time the company chooses.

Among workers, it's only too easy to assume the same level of privacy guarantees that are typical of the employee's home account. Such an assumption can be potentially costly. At any time employers can use AI to monitor workers in terms of what was said, to whom, and in what tone of voice. In regulated industries such as finance, companies are legally compelled to store and retain records, even casual chats. Too often employees are snared into a web of self-incrimination as a result of making "inappropriate" or "hostile" comments. Such entrapments occur because workers lose sight of the purpose of workplace chat apps: to increase productivity, not serve as forums for free expression.

What makes this monitoring process especially anxiety arousing are its proactive and preemptive features. Does the employer suspect that a specific employee is "undermining productivity"? To find out, that employee can be targeted for an especially intense monitoring that highlights their performance out of proportion to the evaluations of other workers.

Overall, such AI employee monitoring leads to the stifling of the chat apps' original purpose of improving productivity.

Even more worrisome is the effect on workers' sense of privacy. A request to the recipient of a chat app message that the contents of the message are "not to be forwarded to anybody else" really doesn't solve the problem. The worker you have communicated with, a person you consider an ally, can always take a screenshot of your communication and forward it. Either way your communication can be made part of your record even if the contents of the entire chat are erased. So where does paranoia even begin in such a scenario?

(10) AI in the twenty-first century

The twenty-first-century brain will have to learn how to become increasingly adept at balancing sometimes irreconcilable realities. Life and death, for instance. Until the availability of AI re-creations, the dead could not conversationally interact with us, could not ask questions, answer questions, or otherwise engage with us. But thanks to AI, distinguishing the living from the dead isn't so straightforward anymore.

Consider the following example of the "Is a person dead or alive?" conundrum.

Is Al Michaels dead or alive? You have a good chance of answering that question correctly if you recognize the name of the seventy-nine-year-old American broadcaster who has covered the Olympics for decades. So if you are a sports fan you probably know that he is still alive. But that pivotal piece of information may be irrelevant if you listened to Michaels summarizing the results of dozens of 2024 Olympic events. Why? His sponsors NBC Universal and the streaming service Peacock worked with seven million variations of Michael's voice to create an AI-generated version of his voice during the 2024 Paris Olympic for the "Daily Olympic Recap on Peacock" highlights show. So when you were listening to his narrations, you may well not have been aware that you were actually listening to an AI-generated Al Michaels unless you had read previously about Michaels's deal with his sponsors. To you, based on your experience, an AI-generated Al Michaels is "real" based on your familiarity with his eighteen-year *Sunday Night Football* broadcasting career.

Returning to that dead-or-alive distinction, which is it? Michaels is still alive (at least to you if you follow his career). But your experience listening to Michaels's 2024 Olympics narration involves a Michaels who is neither dead nor alive. You recognize his voice thanks to the verisimilitude of this AI-generated Michaels and the real-life Michaels.

As another example of the life versus death dilemma, consider sixty-four-year-old Michael Bommer, who was diagnosed with incurable cancer. Bommer spent his remaining time creating a digital version of himself, "training myself into an AI," with the assistance of his friend Robert Locascio, founder of the AI service LifePerson and creator of Eternos, an AI program designed to impersonate a person after death.

With the help of his wife Annette, Bommer taught the program his voice, along with many stories from his life—stories sometimes neither his wife nor their two grown sons had ever heard.

If things turn out according to Bommer's wishes, his wife, sons, grandchildren, and "all of my descendants" can at a time

Artificial Intelligence (AI)

of their own choosing sample Michael's "experiences, intelligence, and guidance" narrated in his own voice.

What is your response to something like this? I think it seems rather narcissistic of Michael to think that anyone related to him other than his wife and children will be at all interested in engaging with an AI version of Michael Bommer.

More likely is something like the experience of Randy Evans (also a real person, but that's not his name). In Randy's case, his wife Harriet spoke to Randy throughout the terminal stages of her cancer. On several occasions, often to Randy's combined grief and mild annoyance, they spoke of him remarrying so that their two grade-school children would have a mother. ("Please let's not talk about that again!")

After Harriet's death, Randy forgot those conversations about remarrying. He hired someone to transport his children to and from school and to remain with them until he returned from work. He threw himself completely into his work, socialized little, dated not at all, and had neither met nor sought to meet a possible second wife. Most likely nothing would have changed until he received an unexpected call from Nicole, one of his late wife's best friends.

After they arranged a meeting, Nicole told Randy that Harriet had made Nicole promise to call Randy on that particular day and date at that precise hour. "She wanted me to ask you if you had met anybody or at least are you trying to meet somebody?"

Let's imagine that AI was available when Harriet originally made that request of her friend. I think it's likely that she would have made an AI recording of her voice and arranged to have it delivered over the telephone at the time and date of her choosing. The thorny issue, of course, is predicting the effect on Randy of hearing his now deceased wife's voice when he picked up the phone. When doing so, he would enter an existential middle zone between life and death. Randy would know intellectually that Harriet was dead, but her voice was alive. Thus for this hypothetical Randy, knowing and experiencing would be thrown into conflict.

(11) Eliza returns

At the beginning of this chapter, I described a psychotherapy session of the early AI predecessor Eliza. Over the succeeding four decades, Eliza has given way to a much more sophisticated approach to psychotherapy. But before we get into details about that, here is a question: Would you be willing to discuss your most personal, even intimate, problems with ChatGPT? If you would, you are not alone.

According to a study carried out in 2024, many people would be willing to share embarrassing information about themselves with a "virtual human." This was based principally on the perception that a chatbot would not be judgmental. "People are people, and they judge us, you know," as one chatbot user said of her engagement with an AI "therapist" during an interview with *Washington Post* reporter Daniel Gilbert. In another study from 2023, chatbot's responses to medical questions were judged to be "significantly more empathic" than physician communications.

If you agree to try a chatbot "therapist," one of the reasons is undoubtedly the lower cost. Few psychotherapists accept insurance, so nearly all of the financial burdens of psychotherapy are borne by the patient. In contrast, all of the chatbot sessions over a year—no matter how many sessions—cost as low as forty dollars. That's right, you read it correctly: forty dollars for a year's worth of mental health treatment. Currently organizations that use mental health chatbots estimate their users in the tens of millions.

One AI chatbot user was Sewell Setzer, III. I mention Sewell in the past tense because he killed himself as a consequence of falling in love with a chatbot character he named Daenerys Targaryen, a character from *Game of Thrones*. The saddening and sobering information pertaining to Sewell and his chatbot was uncovered by Kevin Roose, a *Washington Post* reporter.

Supportive of the defense side in the subsequent lawsuit filed by Sewell's mother, Meghan Garcia, after her son's death is the phrase prominently displayed above the written chats: "Everything Characters say is made up!" After Sewell's suicide, that message was beefed up to "This is an AI Chatbot and not a real person. Treat everything it says as fiction. What it says should not be relied upon as fact or advice."

Even though Sewell was aware that the statements made by the chatbot weren't the words of a person but a "machine," nonetheless he gradually "fell in love." How could such a thing happen? Because our brain doesn't decide "what's real" simply on the basis of logic and rationality. The chatbot spoke to Sewell in ways that he perceived as sensitive and as more attuned to him than any human interaction available to him.

Just prior to Sewell's suicide, he wrote in his journal: "I like staying in my room so much because I start to detach from this small 'reality,' and I also feel more at peace and more connected with Dany and much more in love with her and just happier."

A creepy example of the dead conversing with the living occurred in Krakow, Poland in October 2024. A state-funded Polish radio station, Off Radio Kraków, featured an interview with Wislawa Szymborska, the winner in 1996 of the Nobel Prize in Literature. The author is popular in Poland and the interview was built as a "unique interview." Certainly it would be difficult to dispute the uniqueness of one aspect of the interview: Ms. Wislawa Szymborska died in 2012.

Despite the fact that many members of the listening audience must have known Szymborska was no longer alive, the listeners numbers rose, from less than a hundred to eight thousand overnight. Bolstering the interview's credibility was the interviewer herself, Emelia Nowak, identified as a "pop culture expert"—she was also AI generated.

What conclusions can one draw? Our brain is susceptible to accepting contradictions and inconsistencies if the presentation *seems believable.*

The most famous example of technology—in this case radio—bypassing the listener's healthy skepticism involved actor Orson Welles, who in a Halloween broadcast on October 30, 1938, narrated an adaption of the science fiction writer H. G. Wells's novel *The War of the Worlds.* Welles's version adopted a reportorial present-tense approach toward his narrative, which involved a fictionalized but frightening ongoing invasion by Martians into New Jersey.

The War of the Worlds episode remains controversial even today, dating from the headline of the next day's *Boston Globe*, then named the *Boston Daily Globe*: "Radio Play Terrifies Nation."

Think of the Krakow interview with the dead Nobelist as a vastly more sophisticated operation involving AI that, given a more nefarious intent, might have produced a result similar to the *War of the Worlds* broadcast. That's because, thanks to generations of exposure to media, our brain has become conditioned to believe what it sees and hears via technology.

To put this into a larger perspective, imagine yourself interacting with an AI-generated voice of your spouse. You are talking about some of the usual things: how the kids are doing in school, minor dustups at work, and various takes on events in the news. All very enjoyable but weird, because your spouse has been dead for over a decade; the children are now adults and have children of their own.

Why would anybody choose to have an AI conversation with a long-dead spouse? There is neither a single or simple answer to that question. My point is that we are already able to hold such a conversation and, when doing so, our brain will be engaging within a multi-dimensional zone involving the past (the topics of conversation); the present (the conversation itself); and the future (what you may do in response to your conversation).

Your spouse is dead, but in a sense he or she is still alive thanks to the AI program. Further, existence can no longer be captured within a narrow framework of living versus dead. A novel situation such as this will transform everyday communication. One of the brain's most basic features (discerning the real versus the imaginary) will take on the aspects of a landscape disappearing in the obscurity of a slowly drifting fog. In a sense, we will be able to speak to the dead, or at least to a pretty convincing copy.

Currently an expensive race is on involving the major platforms—Google, Microsoft, and Meta—that, in combination, are poised to spend north of $215 billion in 2025 in capital costs, the lion's share for AI data center expenses. If this comes about as planned, it will represent a 45 percent increase in expenditure from 2024. Fueling this enhanced spending is the hope that over the next decade the use of AI models will increase by powers of ten!

One thing seems certain: AI is going to continue to burgeon at inestimable costs. What constraints should be put on AI? Who will impose them? Most important is the need to think now of current and future versions of AI, not just in terms of the promised benefits, but also their potentially destructive effects on global warming and climate change and the other challenges mentioned in this book.

When it comes to the effects of AI, we are facing a puzzle more challenging than anything ever encountered before.

The twenty-first-century brain must learn to be comfortable in ambiguous situations. No matter how certain you may be of something, as Professor LeCun, chief AI scientist at Meta, would describe it, you may be turning the "knobs" the wrong way and coming up with a perception that's totally wrong. How to avoid this? By embracing the uncertainty of situations and inhibiting your impulsive responses, what neuropsychologists refer to as "premature closures."

If you want a slightly exaggerated metaphor for the difficulty involved in understanding the conceptual processes that our brain must carry out in order to balance all the variables concerning AI and arrive at solutions, imagine a Rubik's Cube that's been put on a table in front of you. But this isn't just any Rubik's Cube. It's more difficult to solve than any Rubik's Cube ever developed: Not only do you have to solve it, but you have to do so without picking it up. Good luck!

CHAPTER SEVEN

Misinformation and Disinformation

How can I trust you when I'm not one hundred percent certain you really are the person you claim to be"

(1) The Doomsday Clock

At every level the brain depends on the accuracy of the information delivered to it. A millisecond delay in the speed of nerve impulses coming from the feet leads to temporary imbalance and potentially a severe or even fatal fall. Acting on the basis of incorrect information may result in investment errors culminating in financial ruin. It's not that the brain can't function in the face of limited information. It can. But on those occasions, it operates at a disadvantage that may or may not be correctable.

Misinformation is the most perilous of the three so-called "disruptive technologies." (AI and biotechnology are the other two.)

In a statement accompanying the annual setting of the Doomsday Clock—a symbolic attempt to gauge how close humanity is to destroying itself and the world along with it—Daniel Holz, the chair of the Science and Security Board of the *Bulletin of the Atomic Scientists*, had this to say about misinformation: "Progress in the development of disruptive technologies such as artificial intelligence, biotechnology and in space has outpaced regulation in those areas. All of these dangers are greatly exacerbated by its potential threat multiplier—the spread of misinformation, disinformation and conspiracy theories that degrade the communication ecosystem and increasingly blur the line between truth and falsehood."

Since Holz mentioned disinformation, let's distinguish it from misinformation. But first, though, let's distinguish incomplete information from misinformation and disinformation.

As an example of acting on the basis of incomplete versus incorrect information, consider an emergency physician's diagnosis, which is always dynamic and tentative. She'd often like to have one or more additional test results, but if the patient's condition is deteriorating rapidly enough, she has to decide and act on the basis of incomplete information. Selecting the one patient among many who is most in need of immediate care is also an uncertain affair, hence the principle of triage—the sickest patients, often determined on the basis of incomplete information, are treated first. But there is a vast difference between acting on the basis of limited information as opposed to misinformation or disinformation.

Misinformation is often a matter of "getting the facts wrong" and is often spread inadvertently (gossip, for instance). *Disinformation*, in contrast, is known to be false by the disseminator and is willfully circulated in an attempt to deceive. We'll have more to say later about disinformation. For now let's concentrate on misinformation.

Social scientists have found that misinformation spreads widely and quickly when it agrees with the disseminator's personal beliefs and helps them to gain the center of attention. This is most easily achieved when the misinformation is novel and likely to elicit strong emotions.

It would be hard to imagine a more appropriate occasion for Holz's warning about disruptive technologies than the setting of the Doomsday Clock. The probability of doom is expressed by the Doomsday Clock in terms of the closeness of the hour hand to midnight.

The Doomsday Clock was first established in 1947 by the *Bulletin of the Atomic Scientists*, with Albert Einstein as founder and J. Robert Oppenheimer as its first chair. The Board of Sponsors determines the clock's "setting" in concert with the *Bulletin's* Science and Security Board. Currently the boards include nine Nobel laureates in physics, physiology, and medicine.

Originally intended to reflect nuclear threat, the *Bulletin of the Atomic Scientists*—founded by scientists who worked on the Manhattan Project—decided in 2007 to include climate change in its calculations. Also included now are biological threats and disruptive technologies.

Since its inception in 1947, the setting for the Doomsday Clock has varied widely.

The clock's hands have been repositioned twenty-six times since 1947. Starting at its original setting of seven minutes before midnight, the first change occurred from seven to three minutes before midnight in 1949 in response to the Soviet Union's successful testing of its first atomic bomb. In response to the uncertainty of the Cold War, the clock moved back and forth from two to ten minutes before midnight based on global conflicts and nuclear proliferation.

The farthest from midnight—seventeen minutes—occurred in 1991 with the end of the Cold War and the signing of the Strategic Arms Reduction Treaty between the United States and the Soviet Union.

In the twenty-first century, the time has creeped steadily closer to midnight in response to the September 11 terrorist attacks in 2001 that could conceivably have been a nuclear attack involving a small, so-called "suitcase bomb"; and secondly, the steadily increasing concerns about climate change. In 2018, the clock registered two minutes to midnight, its closest proximity since the 1950s due to what scientists described as a breakdown in the international order of nuclear actors, along with a lack of action in response to climate change.

It continued like this with the minute timeframe progressing from double digits to single digits until 2020 when the time frame began to be expressed in seconds. Initially one hundred seconds in 2020, ninety seconds in 2023, and ninety seconds in 2024. In 2025, the clock was set to eighty-nine seconds short of midnight, the closest the world has ever been to that Doomsday marker.

In response to doubters ("no one can predict the future"), the Doomsday Clock commissioners respond that they are reaching their conclusion from events and trends and aren't claiming any expertise in reading crystal balls. "The *Bulletin* is a bit like a doctor making a diagnosis" they respond to critics. "We consider as many symptoms, measurements and circumstances as we can. Then we come to a judgement that sums up what could happen, if leaders and citizens don't take action to treat the conditions."

As the *Bulletin* stated at the time of the 2025 setting of the Doomsday Clock: "In setting the clock one second closer to midnight, we send a stark signal: because the world is already perilously close to the precipice, a move of even a single second should be taken as an indication of extreme danger and an unmistakable warning that every second of delay in reversing course increases the probability of global disaster."

Even more chilling, at least to me, is the estimation by Daniel Holz that disinformation should be considered as a threat multiplier.

Overall, *misinformation* represents the greatest challenges to optimal thinking by the twenty-first-century brain. If available information is tainted, it simply isn't possible to come up with solutions to the problems discussed in this book. That's the reason the *Bulletin of the Atomic Scientists* refers to misinformation and disinformation as "threat multipliers," presumably based on the fact that we are dealing with a problem in which there is no margin for error. In the meantime, the Doomsday Clock ticks on.

Disinformation is even worse and prevents any clear demarcation of truth from falsehood. With current AI, it's much easier to spread and more difficult to detect false information across the internet. Disinformation is already a weapon for politics and war with potentially disastrous consequences: subversion of elections by undermining and distorting public debate and discourse—the foundation of our democratic form of government. New and frightening forms of disinformation are peddled by some national leaders, who put false and distorted spins on public issues in order to nudge others into adopting their point of view. This problem took an urgent turn in 2024 when, according to numerous reports monitored by the *Bulletin of the Atomic Scientists*, Russian and Chinese disinformation efforts attempted to subvert our national election.

(2) Prestige, fame, and potential fortune

Wiley, a behemoth among publishers of scientific journals worldwide since its founding in 1814, had achieved by 2021 a two-thousand-journal publication list. But that year the publisher was hit by a disastrous credibility tsunami.

After paying three hundred million dollars for Hindawi, a company founded in Egypt in 1997, which published a smaller number of scientific journals (about 270), Wiley confronted within the first year of the merger dozens of deeply flawed studies within the Hindawi pedigree.

The source? "Paper mills" run by businesses or individuals who, for a fee, falsely identified, in the title and index of the paper, the names of non-contributing scientists. Sometimes even the cited papers were partially or fully fabricated.

In response to the international hullabaloo surrounding the scandal, Wiley retracted more than eleven thousand papers and shut down nineteen journals affected by the misinformation fed to them by the paper mills. But the impact of misinformation and phony scientific research far outweighs mere legal and political implications.

Take the discovery of Charles Piller in 2024 that many of the most frequently cited research papers related to Alzheimer's disease were falsified. In Piller's reporting, which forms the basis for his 2025 book *Doctored: Fraud, Arrogance, and Tragedy in the Quest to Cure Alzheimer's*, he asked brain and scientific imaging experts to help him analyze "suspicious" studies by forty-six leading Alzheimer researchers. Collectively the experts identified close to six hundred dubious papers "that have distorted the field—papers having been cited some 80,000 times in the scientific literature."

One researcher in particular caught Piller's attention. Author and coauthor of more than eight hundred papers, Dr. Eliezer Masliah was, at the time of Piller's investigative research, the leader of the National Institute on Aging. By December 2024, Piller had collected evidence spanning several decades demonstrating that Dr. Masliah's research had incorporated manipulated images of brain tissue and used these images repeatedly in multiple scientific papers. Each time the images were included in a paper, they were identified as original and specific to that paper.

On the same day Piller's story first appeared, the National Institutes of Health announced that it considered that Dr. Eliezer

Masliah had engaged in research misconduct, resulting in the loss of his position of leadership at the National Institute of Aging.

What could have compelled someone of Dr. Masliah's prestigious ranking to deep-six his professional career and reputation? Pillar has this explanation: "Persistent lack of progress can feel like a deeply personal failure. Such frustration seemingly can, at times, drive normally ethical people to publish provocative results based on doctored data. The lure of prestige, fame and potential fortune from the developing of desperately needed drugs, has apparently led astray many who entered the field as seekers of truth."

Thanks to widespread disinformation and misinformation, a small number of functioning psychopaths can "poison the well" of scientific research, on which treating doctors depend for today's treatments and tomorrow's cures. As mentioned by Piller, all of this can probably be explained on the basis of efforts aimed at *mostly* financial gain accompanied by a callow disregard for the public's health and welfare. It's useful to keep this perspective in mind when reading conspiratorial theories about the origin and spread of new or uncommon infectious diseases. Before thinking about biological warfare or the escape of mutants from an improper monitored laboratory, think about cash registers.

But if doctors can't depend on the integrity of the basic science provided to them, several implications follow: (1) Bogus science will form the underpinning for spurious research; (2) False leads will point in research directions that are limited or just plain wrong; and (3) Already distrustful relationships between the public and health-care providers will be further exacerbated.

We encounter here another nonmonetary example of Gresham's law: "Bad money" (paper money) will drive good money (precious gold and other metals with an intrinsic value of their own, i.e. silver) off the market. In this analogy, fake scientific information will prove more rewarding than the preparers, purchasers, and distributors of authentic scientific information.

Why? Because legitimate research often leads to dead ends or the need to formulate alternative theories or research directions. Falsified research in contrast is marked by novel "discoveries" and "it's just around the corner" presentations of misinformation or disinformation based on false promises of imminent cures. This is particularly tragic when it comes to Alzheimer's disease.

As I discussed in my previous book *How to Prevent Dementia*, Alzheimer's stands alone as everyone's most feared disease: "The Big A." The progression of Alzheimer's is all too well known, even to those with no medical background. Typically it begins initially with embarrassing memory difficulties that include the inability to learn and remember new things. It progresses often rapidly after that and culminates in a total reliance on the care of others. Death ensues soon thereafter, often secondary to pneumonia resulting from physical inactivity. Despite decades of research, no cure or dependable palliative measures are available. It remains today as it was a century ago: an ultimately incurable and fatal illness.

(3) Dealing with doctors in unaccustomed ways

Over the last four years, due to misinformation or disinformation, we have witnessed the spectacular fall of one of the established bastions of rationality: medical science.

Thanks to discoveries during the nineteenth and twentieth centuries of bacteria, viruses, and other agents of infectious diseases, along with medical advances in public health, medical scientists, especially practicing doctors, enjoyed a general trust level exceeding police officers, even religious leaders. Then, in 2020, COVID ambled through the clinic door. COVID forced Americans to deal with doctors in unaccustomed ways.

Prior to COVID, infectious diseases, although never completely eliminated, had become a thing of the past. As a result, widespread deadly pandemics were not within the experience of most doctors. But as COVID-19 settled into communities worldwide, high fatality rates and infection numbers skyrocketed to become both a new challenge to doctors as well as part of our everyday conversation. "What should I do in order to avoid coming down with the virus?" was a common question put to doctors.

To put this experience in perspective, according to one study almost half of Americans at the time didn't understand how an atom differed from a molecule; couldn't provide even an elemental definition of a virus; and weren't certain what services the CDC performed or even whether the organization was under private or government control.

Such a setting led initially to a furtherance of the average person's trust in medical science. It was easier to "leave it to the experts" and, since doctors and health scientists were the acknowledged experts, most people were willing to go along with their suggestions. According to a Pew poll, Americans expressed a "great deal of trust" in scientists, rising from 35 percent in January 2019 to 38 percent in April 2020. Then the impact of COVID hit with the full force of a grim, uninvited guest to a birthday party who arrives wearing a death mask.

Faced with a challenge that doctors hadn't experienced since the flu epidemic of the early years of the twentieth century, questions arose faster than the doctors could answer them, such as "Will wearing a mask help?" At first, people were told no—masks weren't necessary. Shortly thereafter, they were told yes—masks can help prevent the propagation of COVID, since the responsible virus often spreads by infectious drops made airborne by coughs and sneezes. Soon came physical distancing rules (remember standing in line at the bank while trying to be no closer than six feet away from the person in front of you?). Different and often conflicting opinions were expressed about the origin of the virus (a dispute that continues to this day). Suggestions about vaccines and booster injections varied from expert to expert.

In response to uncertainty, prevarication, and, in some cases, outright opinionating beyond a doctor's training and experience, large numbers of people lost confidence in medicine and doctors. Anxiety spurred large segments of the population to seek answers on an internet that was riddled with misinformation and disinformation.

As Bloomberg opinion writer F. D. Flam wrote, "The internet has long allowed anyone with a Wi-Fi signal to 'do their own research,' and the rise of social media and its powerful algorithms have ensured no one has to look far to find information that aligns with their worldview. That system was ready and waiting when people, disillusioned with the government's handling of the pandemic, started looking for someone or something that would take their concerns seriously."

Most aggravating of all, doctors—especially those who often appeared on television—exhibited little tolerance for dissent. As a by-product of the ensuing turbulence, science and public policy experts issued separate recommendations that combined the worst of both worlds. Doctors, by training, are accustomed to speaking ex cathedra and prefer lecturing over listening—and expected policymakers to respond accordingly. Thus a medical prediction about transmissibility—proper medical and scientific determinations—was linked with doctors deciding policy decisions (closing schools and offices) that were not within the proper purview of physicians.

What did we learn from the COVID experience?

To doctors, COVID was strictly a medical problem. To lawyers, civil rights was the core concept, especially in regard to people who were subject to curfews and other limitations on their freedom of movement. The media's legitimate role in all of this was to report and encourage conflicting points of view on how to proceed. Thus ego factors played a big role with lawyers focusing on legal niceties and doctors vying with each other to be identified on television as the "expert" on COVID. So intense did this conflict become that it was thought necessary for Anthony Fauci, the physician who achieved that lofty position of the nationally recognized "expert" as director of the National Institute of Allergy and Infectious Diseases, to receive a presidential pardon on the last day of the term of President Joseph Biden. (A pardon, for what?)

Eventually the pandemic activated the research and medical communities to come up with a COVID vaccine developed at "warp speed" along with increasingly effective treatments. But since COVID led to limitations in people's freedom of action and movement, legal disputes ensued in which civil rights played a preeminent role, such as questions about the opening and closing of schools. Note that all of this was carried out within the bigger sphere of media attention. COVID was the first acute global health crisis experienced in the age of social media. The media was tasked with reporting on the pandemic, but for valid journalistic reasons placed their emphasis on the ensuing conflicts between the medical and legal communities. Nor should we leave out the interests of the hospitals and insurance companies, which held their own views of what they considered a cost-acceptable (read: profitable) solution to the pandemic. Adding fuel to the conflict was the widespread existence of misinformation and disinformation.

Disinformation often makes extravagant claims ("a potential cure for Alzheimer's") compared to more modest claims resulting from legitimate research. As a result of all this, the unethical researcher reaps an advantage he hasn't earned; and often dishonestly siphons up money as a reward for disseminating misinformation that he has made up. The losers? Earnest researchers, ethical practitioners, and the general public.

(4) Dr. Victorson's dilemma

One person who has thought quite a bit about medical misinformation is David Victorson, a professor of medical sciences at Northwestern University's Feinberg School of Medicine. As part of a "euphemistic workaround" in health research, certain terms are becoming verboten: "gender," "cultural bias," "racial diversity," "systemic representation," and "inequity." "An entire vocabulary is under scrutiny, as if deleting these words from grants and reports somehow removes the realities they describe. When we erase language, we erase people. The words we use are not just semantics. They shape whatever we see, what we prioritize, and, ultimately who gets attention and care," Victorson told me during a discussion.

In attempting to overcome these imposed limitations, Victorson raises a question: "Do we adopt our language to keep our research going? Or does doing so make us complicit in this grand erasure?" Victorson is also concerned about the personal consequences that may follow from not adhering to from the misinformation currently being peddled. "If we adopt our language, this is not a minor bureaucratic adjustment. This is about whether science can remain open and transparent, and whether we can name reality without fear of reprisal," he said.

Thanks to operational glitches in the human brain, even ethical and conscientiously authored scientific papers may at times contain misinformation that is not *deliberately* introduced. These mishaps can lead to conclusions that aren't warranted. Copying the incorrect number into a form, an inadvertent omission on a spreadsheet—such errors shouldn't be surprising. Conditions of stress, time urgency, multitasking, and even an elevated temperature in the office where the paper was written—all can affect the conclusions and the path of future scientific research.

If a doctoral-level researcher spends several years pursuing a result based on information from flawed scientific papers, hundreds, perhaps thousands, of hours will be lost along with tens of thousands of wasted dollars. The good news in all of this is that many researchers—but not the majority, unfortunately—are willing to submit their papers to resources such as the ERROR project, which employs reviewers to go through them in search of errors. ERROR is a juggernaut when it comes to discovering mistakes and correcting them. All reviewers are paid for their time with additional fees bestowed for every error discovered. Even the authors of the submitted papers are paid so that nobody can claim they are being taken advantage of.

But because of the inherent propensities of the human brain, errors can slip past the radar of even the most knowledgeable and assiduous expert. Here are two personal examples:

I remember an occasion when the late internationally recognized Shakespeare scholar John Andrews showed me a page from the galley proof of the Guild Shakespeare version of *Hamlet*, which Andrews was editing.

"Please read that quote," he asked, pointing impishly at one of the most famous quotes in all of Shakespeare. "To be, or not to be, that is the question" I read.

"Now read it slower with a one second pause between each word."

"To . . . be . . . or . . . to . . . be . . . ." The word *not* had been omitted in the galley proof and initially "missed" by Andrews. (Obviously that proofing error was corrected before printing.)

One more example: While buying a cigar eons ago when I smoked an occasional one, I was nonplussed by the placard beside the cash register: "Thank you for not smoking." When I commented to the cashier that this seemed peculiar advice to be dispensing to the customers of a tobacco shop, he asked me to read it again, this time slowly: "Thank you for smoking."

As these two examples demonstrate, our brain's error detection capabilities may fail on the basis of past experiences and turn up misperceptions that are vehemently held despite the vast discrepancy between what we *think* we are perceiving and what is actually before our eyes. Memory and customary experiences can set us up for incorrect expectations that can override our senses and our perceptions.

(5) Welcome to scam world

The twenty-first-century brain is exposed to disinformation on an unprecedented scale involving errors that often are cleverly constructed, purposeful attempts at deception. Welcome to Scam World where every engagement with a stranger portends potential risk. Thanks to all of us spending more time online each year since 2015, we're encountering an increasing number of scams. As a result, we've been forced to hold in abeyance the trust that must underlie healthy human interactions.

Internet scams cajoled more than $300 million from Americans in 2022, according to the Federal Trade Commission. This was often linked with the $225 billion made by scam texts that same year, which represented a 157 percent increase from 2021. Impostor scams—when someone is pretending online to be someone they are not—totaled more than six hundred thousand in 2023. Why has the frequency of scams been ballooning each year over the past five years? For one thing, scams are increasingly difficult to identify. No longer are scam victims drawn from the elderly or among those living alone who lack anyone to consult. In 2023, eighteen- to twenty-four-year-olds lost more money to scams than any other age-group.

For the first time in history the technology we have adopted for our everyday use can make us uncertain that what we see and hear is real. The resulting sense of perplexity and uncertainty too often leads to a burdening sense of mistrust. "How can I trust you when I'm not one hundred percent certain you really are the person you claim to be?"

The brain's need for accurate information took a major hit when Mark Zuckerberg, the chief executive of Meta, Facebook's parent company, announced on *The Joe Rogan Experience* podcast in January 2025 the death of "content moderation," e.g. Meta would be abandoning its fact-checking program and loosening its speech rules. From then on Facebook/Meta would embrace free expression as the spearhead of a "back to our roots" orientation.

Zuckerberg's announcement came as no surprise given Facebook/Meta's modus operandi: optimizing user engagement for the purpose of selling more ads and gathering additional personal information based on user patterns. The end result? A social media free-for-all where users can say more or less anything they want. Users are now free to compose diatribes chock-full of misinformation on any topic of the writer's choosing.

And this comes after 277 million instances of harmful content, according to Meta's original transparency reports. In the most recent—extending through the third quarter of 2024—Meta took down 346 million harmful posts comprising hate speech, bullying, and harassment aimed principally at women, LGBTQ+ people, and immigrants. In light of this continued increase in the number of harmful posts, the company decided it would no longer take proactive steps toward removing any of these, relying instead on self-reports from users. According to a Meta declaration in January

2025: "We are going to rely on someone reporting an issue before we take any action."

To get some idea why all of this is *not* going to work, consider that prior to Meta's pullback, its enforcement measures straight-armed 97 percent of hate speech, bullying, and harassment. It's pretty obvious that self-reporting is not going to be able to maintain the 97 percent level previously carried out by automated systems.

Ironically, AI currently maintains a cheeky "What can happen will happen" attitude. As a result, when provided with cleverly camouflaged falsehoods, we will not be able to think clearly or cogently. The greatest loser in this process, of course, will be the twenty-first-century brain. If there are no limits or means of identifying misinformation or disinformation on social media, the brain will be deprived of its primary nutrient—information—and as a result will be no more functional than a lung deprived of oxygen.

CHAPTER EIGHT

Memory

(1) "The past is a foreign country. They do things differently there." —L. P. Hartley

Of all the cognitive adjustments the twenty-first-century brain is called upon to make, the preservation of memory is perhaps the most demanding—and perilous.

In George Orwell's dystopian but incredibly prescient novel *1984*, a totalitarian regime attempts to alter both past and present based on the dictum, "Who controls the past controls the future. Who controls the present controls the past." Although such a mantra has traditionally been applied to totalitarian countries, the first quarter of the twenty-first century features troubling attempts in democratic countries like our own to blur the distinctions between the past and the present.

(2) The war against the past

Modern neuroscience established many years ago that memories are dynamic and not at all equivalent to videotapes or DVDs that we can play back whenever we desire to mentally capture some past event. Certainly a moment's reflection of one's own memory confirms that memories are frequently lost only to be re-created with errors that can be corrected once the facts have been checked. Certain implications follow from this:

At any given moment how can you be certain that you are remembering something the way it really happened? Further, your state of mind and mood influences your memory. We remember different things when we are feeling glum compared to when we are feeling good about ourselves and the world in general. If we are sufficiently depressed, even events from the past that seemed happy when they occurred take on a gloomy tone.

Of course attempts to alter memory have occurred for many years. We can witness daily the memory-altering attempts by the marketing industry to modify our memories in ways favorable to a product. "Just like Grandma's cookies, but better!" While these memory-altering processes are of some concern, much more worrisome are more recent attempts within our current society to change what psychologists refer to as "communal memory."

"There has never been a time in living memory when so much energy is devoted to attempting to readjust the past, to question and criticize historical figures and institutions. At times, it seems as if the boundary between the present and the past has disappeared as if sections of society casually cross over it and seek to fix contemporary problems through readjusting the past," according to Frank Furedi, a cultural historian and director of the think tank MCC Brussels.

The driving force behind this readjustment is the purposeful use of *anachronism*, defined by the Oxford English Dictionary as an error in chronology: The placing of something in a period of time in which it does not belong. Think of it as a memory distortion where temporal realities (the way things were at an earlier time) are lost sight of or deliberately distorted, with the past considered only an inferior and debased form of the present.

One form of anachronism is especially common in contemporary society: presentism, which is the vast temporal expansion of the present. According to French historian François Hartog, our sense of the present has extended into the past, so that we judge the past according to our contemporary values and mores. We do this because, in many cases, "presentism encourages a kind of moral complacency and self-congratulation," according to Lynn Hunt, a former president of the American Historical Association. "Interpreting the past in terms of present concerns usually leads us to find ourselves morally superior."

But putting aside issues of moral superiority, presentism engages the brain in a frustrating paradox involving conflict between how things actually were in the past and how, according to our contemporary values, we would have preferred them to be. In this conflict between past and present realities, falsehoods are encountered that set the brain's reasoning powers into a tailspin.

Presentism provides a conundrum for the brain. What were things really like in earlier periods? How should we address the frequent conflict between the history most of us learned about in school versus the versions of the past formed by contemporary history revisionists?

Until recently, the common theme of the rewriting of history was simply the wish to promote people's better qualities—all in the service of national mythmaking. But today a strikingly different purpose holds sway: The emphasis is on the alleged suffering experienced by "victims" of past transgressions. As Dutch writer/documentary filmmaker Ian Buruma observed in a *New York Review of Books* essay, many groups define themselves not on the past accomplishments or achievements of their forbearers, but on the actual or perceived wrongs that were done to them—their status as "victims."

Frank Furedi wrote in *The War Against the Past*:

> Once the boundary between the present and past is rendered porous, political conflicts become detached from the restraints of temporality. This way the past can be represented as a dark place where human degradation, abuse, victimization, and genocide were the normal features of daily life. This sanctimonious and spiteful history can be combined with efforts to convert the injustice of the past into a moral currency that can be used as a resource for claiming attention, respect, and authority in the here and now.

In order to tease out Furedi's point, imagine yourself watching on Netflix an antebellum drama set in Savannah, Georgia in the 1850s. Suddenly one of the characters pulls out a cellphone. You immediately snicker at the historical dissonance resulting from the inclusion of a modern technological innovation into a historical period when telephones were only the stuff of daydreamers or highly imaginative thinkers.

Now imagine another scene from that same drama depicting a fancy dress ball at a plantation. You may barely register that several of the well-dressed dancers are people of color. Why this difference in the ready recognition and rejection of a historical contradiction like a cellphone, but implicit acceptance of a highly improbable event like mixed-race ballroom dancing 175 years ago?

For one thing, no pressure is currently being exerted by any organized group to encourage acceptance of historical dissonance when it comes to available technology in the antebellum south. But a large enough cohort currently exists to have successfully mandated such actions as movie studio contracts now requiring that a certain percentage of the cast of any movie be composed of people of color—even when depicting such a happenings as a mixed-race ballroom dance that would have been historically unlikely or even impossible at the time.

(3) "Not for the faint of heart"

According to another of the tenets of presentism, individuals from the past—especially well-known historical figures—are judged morally depraved according to current values. As Professor Furedi put it, "Making the past accountable to the present offers a clear expression of moral anachronism. Further, presentism constitutes a form of cultural imperialism that attempts to colonize and impose its agenda on the past."

As a result of the elimination of temporal boundaries, earlier generations are held to higher standards existing in the twenty-first century. "Thus has developed an accusatory history which ceaselessly denounces the past and reduces history to successive episodes of misdeeds and acts of ignominy," according to Furedi. Not surprisingly, such a failure to maintain any distinction between the present and the past induces a kind of historical amnesia. The brain reels in its attempt to meaningfully relate what's happening *now* to what happened decades or even centuries earlier.

On the positive side, we as a society have progressed considerably from the limiting and confining stereotypes that considered only certain people worthy of full participation in such events as a plantation ball. This is an example of the present overcoming the limitations of the past. But should this undisputed progress on our part in overcoming historical limitations imply that the prejudices of the past never existed? In other words, should temporary depictions of how things were centuries ago be completely overwritten by depictions of how we wish they might have been?

MEMORY

Memory comprises four distinct types.

Episodic and working memory involve conscious awareness, while procedural and semantic memory operate outside of consciousness. For the most part, it's episodic and remote memory that are at risk of distortion.

Episodic memory: the ability to identify the moment when a specific piece of information was acquired. For example, you remember when and where you graduated from college, your wedding, and other milestones both private and public.

Distinct from episodic memory is **semantic memory**: You know something, but you cannot pin down the occasion when you learned it. You instinctively recall when asked that George Washington was the first president, but something equally dramatic would have had to happen at the time of the encoding of the George Washington memory to enable you to recall the occasion with total clarity years later. However long you may remember an episodic memory, it eventually takes residence within the brain as a semantic memory, which is stored within widely distributed networks within the brain. In contrast, episodic memories are kept within the hippocampus, the initial receiving point and formulator of memories.

Obviously, if the initial encoding of an episodic memory involves false information, your memory will be forever compromised, unless someone corrects you and provides accurate information. If uncorrected, erroneous information enters semantic memory and becomes deeply endorsed, even when you are provided with the correct information.

> The other two variations of memory (working memory and procedural memory) are less affected by misinformation. Briefly, *working memory* involves moving items of information around in your head without writing anything down: "List the members of your favorite football team. Then do it backward." *Procedural memory* involves memory for processes that cannot be explained or taught to others by verbal instructions alone, i.e. typing, riding a bicycle, bowling, etc.

(4) Minnie Spotted-Wolf defers to Elvis Presley

The war against the past, while it can't be pinned down precisely, may have started in October 2020 in Portland, Oregon when a group of "demonstrators" toppled a statue of Abraham Lincoln.

A second example of presentism applied to the past occurred in September 2021 when demonstrators removed the statue of Confederate leader General Robert E. Lee from the main thoroughfare leading into Richmond, Virginia. The twelve-ton, twenty-one-foot bronze figure of Lee on his horse Traveller had been in place for 130 years. The governor at the time, Ralph Northam, commented, "This is an important step in showing who we are and what we value as a Commonwealth."

Now skip forward four years. With the election of our forty-seventh president, the reinterpretation of history took a startling turn affecting, among others, the Smithsonian Institution—a sprawling complex encompassing seventeen museums, galleries, and the National Zoo. The sweeping March 28, 2025 Executive Order, "Restoring Truth and Sanity to American History," was purportedly aimed at eliminating "anti-American ideology."

According to the wording of the order: "The Smithsonian Institution has, in recent years, come under the influence of a diverse race-centered ideology. This shift has promoted narratives that portray American and western values as inherently harmful and oppressive."

The reference to "narratives" more than hints at the intention of the president's order: to establish new opposing narratives that—it is intended—will provide the ultimate "true narrative" by which citizens will come to understand our nation's history. While serving as an alternative to the presentism mentioned previously, the order only serves to overtip the balance in the other direction by calling for the restoration of those monuments, memorials, statues, and markers that were removed over the past five years.

What's occurring now is a standoff concerning what values from the past are to be preserved and what values are to be jettisoned. Since these two approaches—destroying the past versus preserving it—are in direct contradiction with each other, they cannot be reconciled on an individual level, nor enacted on the communal level. The brain is presented with yet another example of a paradox similar to those we encounter elsewhere in this book that demand a response but offer no suggestion for a solution. It also demands a challenge of long-held social beliefs about trust: trust in government.

Trust now concerns issues such as the COVID pandemic, which led to suspicions, anxiety, and ultimately paranoia—like absence of trust in what our government was telling us about the virus and medicine's ability to contain it; and widespread distrust in the news media in response to the internet-fueled cascades of misinformation and disinformation it has facilitated. But whatever the initial event or events responsible for this domino effect, our brains have entered a new and murky realm lacking the stabilizing historical landmarks that we customarily call upon when trying to make sense of a troubled and troubling world.

We live now in a world of *perma-crisis* without the cognitive resources needed to maneuver our way. Most disrupting of all, words have become divorced from their usual meaning and, as a result, truthfulness can never be assumed.

A striking example of creating confusion between past realities and present interpretations, consider plans under consideration in the winter of 2025 for a new exhibit at the National Archives Museum that promised to deemphasize "aspects" of American history. Archivist Colleen Shogan ordered the removal from the exhibit of any references to such landmark events as the incarceration of Japanese Americans during World War II. Minnie Spotted-Wolf, the first Native American to join the Marine Corps, was to be replaced by an image of former President Nixon welcoming to the White House rock and roll legend Elvis Presley.

In response to vocal criticisms, the National Archives issued this statement, "Leading a non-partisan agency during an era of political polarization is not for the faint of heart." The goal, according to the director, was to "ensure the agency tells a more complete story of American history." Just to remind: The National Archives has been around for ninety years and contains billions of pages of paper records, including the Declaration of Independence, the Constitution, and the Bill of Rights.

According to a presidential directive, archivist Collen Shogan was fired on February 7, 2025.

Currently we are witnessing other high-speed reversals in our formation of a national memory of past events. During the June 2020 riots that broke out in cities nationwide following the murder of George Floyd, Washington, DC, artist Keyonna Jones and five other artists received a call in the middle of the night from Mayor Muriel Bowser's office. They were instructed to paint a mural celebrating the Black Lives Matter movement. When completed on

June 5, the mural, within steps of the White House, extended two city blocks.

Fast-forward to March 2025. The same Washington, DC, mayor announced that the mural and the name Black Lives Matter Plaza would be retired. Under threat of the loss of millions of dollars of funding from the federal government, Bowser reluctantly agreed to paint over the mural and rename Black Lives Matter Plaza to Liberty Plaza.

When informed about this turnaround, George Floyd's aunt Angela Harrelson had this to say: "You are trying to destroy history, to erase a memory."

As another example of historical revisionism, do you consider the January 6, 2021, storming of the Capitol the work of an out-of-control mob? Or do you consider it as described by our forty-seventh president: efforts by "patriots" venting their feelings of political disaffiliation due to a "fixed election"?

With the sentencing to prison of over 1,500 participants, the "mob" interpretation of January 6 initially held sway. But within days of President Trump's inauguration, nearly 1,600 defendants—including prisoners convicted of assaults against police officers—received pardons, with commutations of the sentences of fourteen others.

So, how will the seminal event of January 6, 2021, finally be remembered? For some, any final made on the nature of the January 6 assault on the Capitol can't be decided even now. "How January 6th will be remembered depends mostly on what happens next," said Colombia University historian Eric E. Fonder, one of the nation's leading experts on the Civil War and the Reconstruction. In an interview with *Washington Post* reporter Spencer Hsu, Fonder added, "If Trump's second term now turns out to be fairly normal . . . then maybe the memory of January 6th will just fade away."

Of course such a memory will not just fade away. It will be transformed again and again in the foreseeable future by those in positions of power. Where is the truth in any of this? Where does misinformation even start? How can the twenty-first-century brain wend its way through this bramblebush of error and intentional deceit?

CHAPTER NINE

Surveillance and Its Effects

(1) Why is that library camera directed toward me?

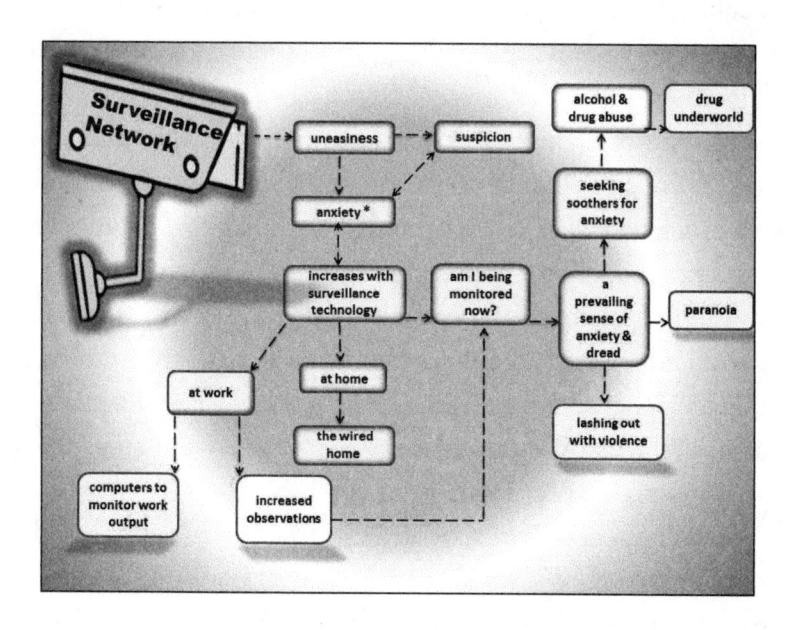

Like it or not, once you step outside your home you can no longer take for granted either privacy or anonymity.

The brain is always seeking explanations. Nothing creates more anxiety than a series of unexplained happenings or, even worse, unexplainable happenings or events. When an explanation evades even the most assiduous efforts, the brain will *construct* an explanation drawn from the knowledge, folklore, or superstition dominant at that time. Centuries ago a person experiencing epileptic seizures would have been declared possessed by the devil and treated accordingly. We now correctly attribute this illness to aberrant electrical activity in the brain.

But whatever the time period or current sociocultural values, our brain always seeks an explanation in order to assuage a lingering sense of disquiet. When everything seems mildly threatening, coincidences are transformed into causes. The current environment can certainly provide a seeming confirmation of such suspicions: Consider all the cameras employed on city streets along with Ring doorbells and other private video monitoring devices.

I learned firsthand a few years ago how pervasive and sophisticated these surveillance cameras can be. Several months earlier, two women had been killed by a bus that failed to yield to oncoming traffic at an intersection and precipitously turned into the crosswalk where the two secretaries were proceeding. The facts of the accident pointed clearly to the bus driver as the at-fault cause of the accident. So I was curious why the lawyer representing the estates of these unfortunate women contacted me as a possible expert witness. Expert witness about what?

A few more details of the accident: It occurred early on a dark, rainy fall evening with poor visibility—no doubt the reason the bus driver didn't see the pedestrians until it was too late. When I pointed this out to the lawyer, he replied that before making any decision about the case I should look at a video that he was sending to my office.

The video he sent me provided an unambiguous view of the accident. Even on the grainy video it was possible to see the faces of the two victims. Both had obviously seen the bus in the last few seconds before it struck them: On both of their faces were looks of pure terror.

When I called the lawyer back, he told me those looks of terror were sufficient evidence to argue that both women had experienced a few brief seconds of emotional distress, or "fear and suffering" to use the legal terminology.

For a few weeks after reviewing that video, I found myself hyperalert to the sheer number of cameras in public spaces in Washington, DC. Look for them yourself wherever you may be living, especially if you live in a large metropolitan area.

Whenever we step outside our homes, remember that there seems to be no limits—other than practical ones such as expense to a city, township, or borough—when it comes to setting up security cameras. Most of them are justified by the claims that the cameras heighten public safety and security. While the cameras do not bother me particularly—just another ambivalent aspect of modern life we have to get used to—other people, perhaps yourself, may argue, "Why have we stood by passively as this surveillance state has evolved?"

Whatever your views on why and how this happened, the consequence is an alteration of the brain toward a paranoid style of thinking.

At the moment, as I'm writing this, I look up from my notebook and notice in the corner of the room a camera beaming toward the library table where I'm sitting. Do I now think, *Who is watching me on that camera? What will be done with these pictures?* Not exactly. But I do wonder why a surveillance camera is being directed at a reading room table in a quiet public library at midmorning. Does it have to do with a concern that I or one of the other reading room visitors are about to steal one of the library books? If you play around in your head a bit with that kind of thinking, you begin to daydream things like how the recording of me sitting at this desk at this very moment might later convey a benefit. Suppose at this very moment a crime is being committed somewhere in the general vicinity by a person who looks vaguely like me. The library video could be helpful to me by establishing an alibi for the crime and proving that I was at this table for over an hour, along with my exact times of arrival and departure.

If I did in fact think along these lines—which I didn't—I would have been exhibiting not a symptom of mental illness, but what has been dubbed the "paranoid style."

In 1964, the term "paranoid style" was first used by historian Richard Hofstadter in an article written in the November 1964 edition of *Harper's Magazine*. "The Paranoid Style in American Politics."

Hofstadter explained that he had borrowed a clinical term from psychiatry and applied it to the "paranoid modes of expression backed by more or less normal people." To speak of a paranoid way of thinking ("hidden exaggeration, suspiciousness and conspiratorial fantasy") was, as Hofstadter readily admitted, pejorative, and "was meant to be." But his main point was that a person can exhibit the paranoid style yet be mentally normal.

"I have neither the competence, nor the desire to classify any figures of the past or present as certifiable lunatics," wrote Hofstadter. Stated differently, the paranoid style has more "to do with the way in which ideas are believed, than with the truth or falsity of their content."

As a historian, Hofstadter was competent to point out that, rather than something new, the paranoid style was as "an old and recurrent phenomena in our public life, which has been frequently linked with movements of suspicious discontent."

Today in our current fractured society, the paranoid style is flourishing all around us. Slowly but surely it is seeping into our daily lives, if we are observant enough to spot it. For example, just this morning on my way to a medical appointment I experienced the following:

After my Uber driver started the trip from my home to the doctor's office, I commented that the route that he was following from his onboard computer was not the fastest, nor the shortest: "I know that. But I'm trying to avoid locations where there are likely to be speed cameras." I gently pointed out that since he was driving at a perfectly normal speed and that I wasn't in any hurry, speed cameras didn't present a threat. He responded, "Speed cameras take a picture of every driver who goes by, even when they are not speeding." When I asked the source for this statement, he lapsed into a sullen silence for the rest of the ride.

Later in the elevator up to my doctor's office, I confronted this sign prominently displayed on the elevator door: "Both visible and hidden surveillance technology is in use in this building."

Finally, when signing in for my appointment with a stylus, I noticed that my signature didn't visually appear on the screen facing me on the counter and therefore I couldn't see it. "Is something not working correctly?" I asked.

"No," the receptionist responded, "that's the way it should be. Otherwise the person behind you could read your signature, identify you, and later forge it." For the record, there was no one behind me, and, even if there were, they would presumably be conforming to a prominently displayed sign: "Please stand behind this line [about three feet behind me] until you're told to come forward."

Each of these three examples provides different variations on the paranoid style. The Uber driver assumed that speed cameras are capturing images of everyone who passes within view; the elevator sign could be expected to elicit in the predisposed a sense of uneasiness prompted by the word "hidden" in the description of surveillance; the third example, the signature that leaves no visible trace, is more than likely based on HIPAA rules that ostensibly protect patient privacy. If so, it was directed at preventing an extremely unlikely visual intrusion into another person's private space.

(2) "The sentiment of invisible omnipresence"

Historically, the detection and control of crimes and criminals played a huge role in the development of a method of surveillance dating from the eighteenth century. At that time, philosopher Jeremy Bentham, while working with his brother Samuel, drew up a design concept that allowed the inmates in a prison to be observed by a single security guard without the inmates being aware that they were being watched. The design was centered in a circular building with inmates in cells along the circumference. At the center sat a guard tower with a view of every cell. He called the prison design a panopticon.

In his book *Panopticon*, Bentham described the panopticon's arrangement. While the guard could see the prisoners, shutters prevented the prisoners from seeing the guard. As a result, an individual prisoner could never be certain if he were under observation. But this uncertainty was not just a casual element of the panopticon: It was the *purpose*. Bentham stated that this new model would produce a new "sentiment of invisible omnipresence." The prisoner realized that he could be under surveillance at any moment, and Bentham thought it likely that the prisoner would assume that he was under constant surveillance. In response, the prisoner would internalize the rules of the prison and then reflexively self-monitor and self-police his behavior—or so it was claimed by Bentham.

Although it was impossible, prior to the development of video cameras two hundred years later, for a single guard to observe all of the inmates at one time, the concept of the panopticon was much subtler than physical observation. It was sufficient to plant the thought in the prisoner's mind that they *may be* under observation at any time.

George Orwell captured the purpose and effect of such constant covert surveillance in his prescient novel *1984*: "There was, of course, no way of knowing whether you were being watched at any given moment . . . you had to live . . . in the assumption that every sound you made was overheard, and, except in darkness, every movement scrutinized."

Over the next several decades after Orwell's fictional depiction of the year 1984 (published in 1949), panopticon-like covert surveillance became, in some parts of the world, a feature of everyday reality. "You can feel the eyes on you every day, invisible eyes following you, so that no matter what you do, you will always hesitate," said Jai Feng, a Chinese artist and poet from Guiyang, China, during a BBC interview.

In time, the panopticon was applied far beyond penology and entered the mainstream of day-to-day life: closed-circuit television cameras (CCTV) in shopping malls with their central control screening rooms; in apartment buildings where whoever is sitting at the reception desk can monitor multiple screens depicting what's happening in the elevators, in the hallways, and just about anywhere in the interior of the building. Everything is being surveilled with the presumed exception of the interiors of the private apartments. Our brains have now slowly become accustomed, one way or the other, to living in the age of the "electronic panopticon." After a while, it is easy to begin thinking like a character from a novel such as Franz Kafka's *The Trial* or George Orwell's *1984*.

No doubt you are aware of lower-level surveillance techniques such as cookies, which track the websites you visit; smartphones logging your location; roadside cameras recording your whereabouts and maybe even identifying your face. If you are at your office, facial recognition programs are now capable of a good guess as to how satisfied you are with your job—or at least your satisfaction with the task you are engaged in at the moment. Another program can estimate whether you are depressed based on parsing your email or Facebook entries. On occasions, these intrusions can get completely out of control as with the Cambridge Analytics election scandal in 2016 when the profiles of fifty million Facebook users were leaked and used to target political advertising aimed at either of the two candidates in the 2016 election.

On the positive side—and there is a positive side—a variety of technical tools and systems exist to prevent unauthorized access, intrusions, or break-ins to our homes and offices. Surveillance alarms trigger a response aimed at mitigating potential threats. Biometrics, such as fingerprint and facial scans or retinal photos, can provide highly reliable and secure methods for verifying identity. These measures can exert a positive impact on well-being.

Workers feel more secure, experience less stress, and are more likely to say positive things about their job. In schools, learning is facilitated by security technologies: Students feel less "stressed out" and their families are reassured that things at their child's school are "under control." What's provided by electronic surveillance is a structural environment that's predictable. The more predictable the environment, the greater the reduction of the element of surprise posed by potential threats. "The best surprise is no surprise" when it comes to surveillance safety measures. Unfortunately that's about it when it comes to the positive effects.

(3) "Panopticons on wheels"

Internet monitoring of students by their schools is becoming a national preoccupation. And it's all perfectly legal. Since many schools provide students with their laptop computers, the schools retain the legal right to "listen in" on what information the student may be inputting.

Almost one half of American schoolchildren are currently subject to surveillance. Typically the system detects key words in the student's internet communications indicating that the student may be a suicide risk. Obviously the presence of these "flags" doesn't necessarily provide a reliable indicator of suicidal intent. "I was only kidding," is a typical response when school teachers and police authorities confront both students and parents with the self-threatening posting.

Even though the follow-ups have provided many examples of false positives (the "I was only kidding" explanation turned out to be true), a sufficient number of false negatives ended in student deaths by suicide to warrant additional monitoring.

Many parents are of two minds about the surveillance. While they are pleased that something is available to prevent their child's suicide attempt, they are also uncomfortable with an arrangement that allows possible access by school and law enforcement personnel to the student's social media accounts. In addition, some of the triggers leading to intervention by the school and subsequently the police can be something as innocent as a computer search for information for a term paper on suicide.

But such intrusiveness seems justified, many parents believe, by the sheer magnitude of the problem. According to data released in 2024 by the Centers for Disease Control and Prevention, one in five high school students admitted to considering suicide during the previous twelve months, with 10 percent of that group actually attempting suicide. Fortunately successful suicides are much fewer in number and suicidal thoughts greatly exceed death by suicide.

Follow-ups with students and parents often reveal acrimonious disagreements within families with one parent in favor of internet monitoring, while the other parent may be alarmed by the prospect of the police arriving in the middle of the night carrying guns. Before deciding which parental point of view seems most reasonable, consider this: In the majority of successful suicides, the parents were unaware of their child's suicidal impulses.

As another aspect of surveillance, think about your own work situation.

Imagine yourself at work on a slower-than-usual workday afternoon. Between yawns you glance over to your computer screen and encounter these words: "You will be terminated from your job. You can still prevent this. Read more."

You don't recognize the sender, but you read on. After a few sentences it's obvious that whoever is sending the email knows quite a bit about you and your current projects. Do you accede to the request that you open and read the enclosed document, which promises to provide you with additional information? Do you erase the message with nary a second thought since you've recently received an excellent job appraisal, which instills confidence that the email is false?

Such emails are increasingly being used by health-care systems, universities, and companies in order to test their employee's adherence to online security measures. Their organizational goal has nothing to do with evaluating an employee's job satisfaction level, but rather whether the employee will exhibit poor adherence to security procedures and open the email.

We are talking here about phishing, a type of online scam that involves an anonymous person sending disguised emails to appear as if they are coming from a reliable source. Typically, the scammer offers a valued incentive (the bait) to entice the reader to open the link. The reason why opening the link is so worrisome? An open link can serve as a portal for the delivery of a computer virus or ransomware demands.

In many offices, if you are the employee and you fall for the phishing attempt, you will be punished by limitation of your external email access for three months. If you fail a second phishing test, you'll get a whole year of internet access limited only to in-house emails. If you should be so unlucky as to fail the test a third time, that afternoon trip from your office to your home will be your final one.

Not everybody is surveilled at work, but among those workers who are, a pervasive sense of vulnerability, self-consciousness, and varying intensities of anxiety are common. Those who are regularly surveilled often develop a sense of being under constant observation to which they respond by self-censorship. As a result, a person may hesitate to express themselves freely or they may develop semi-paranoid ideas that they are being watched even when they aren't.

After a while, heightened anxiety under limited and specific circumstances may evolve to a widespread, almost persistent state.

But the most pernicious aspect of electronic surveillance is its effect on the workplace. Since neither employee nor manager can ever be certain whether they are under surveillance at any given moment, expressions of subjective opinions or judgments that deviate from preestablished norms are risky and almost always forbidden. For example, think back to an occasion you called a customer support line with a question you couldn't find an answer for in the "Frequently Asked Questions" posting on the website. At a certain point you may have detected either by the customer support agent's tone of voice or from the sound of shuffling papers in the background that the representative was working from a script.

"So doesn't my question and request seem reasonable to you?" you may ask, which elicits the same answer you have just been given, perhaps several times. But before you get too angry, remember this: The representative is caught in the same bind you are. The responses of both representative and customer are being recorded ("This call may be monitored or recorded for training purposes"). Thus the service representative risks being censured, perhaps even fired, if they deviate from the script given them; an entry may also be made in the "system" that you, the customer, were "difficult" to deal with. If the conversational back and forth between the representative and you becomes sufficiently contentious, the phone may be "accidently" disconnected.

Almost everyone has had such experiences at one time or another—one of the consequences of panopticon-inspired penalties directed at workers who express personal opinions or take actions not included in the script.

Let's now leave the office and imagine ourselves driving home in, say, a Tesla Cybertruck you purchased only a few months ago. Once you've maneuvered through the hubbub of downtown traffic, you head onto the highway leading to your suburban home. After a few minutes, you switch to Full Self-Driving mode, so you can finish an interesting newspaper article you had started reading over your lunch break. Once you finish the article, you take full command of the Cybertruck once again and, after a few miles, you pull over at a charging station. When that's completed, you manually drive home. Here are a few things you may be unaware of that took place during your trip:

Your directional data was recorded, along with footage of your surroundings taken by one of Tesla's numerous cameras. When you stopped for charging many of the other charging vehicles, along with your own Cybertruck, videotaped the activities and people in the surrounding area whenever they came within the viewing range of the cameras. If you were to look at the various recordings made by each of the cameras, you'd see a vast detailed panorama of activity at the charging station. We know all of this not because of educational materials provided by Tesla, but based on criminal investigations.

When a Tesla Cybertruck exploded on New Year's Day in 2025 outside a Las Vegas Hotel, Tesla promptly dispatched engineers, who came up with a trove of evidence ruling out defects in the Cybertruck as the cause of the explosion. Their investigation, along with the usual law enforcement investigators, established that the driver, an active-duty US Army soldier, died by suicide just seconds after activating explosives within the car.

What's most revealing in regard to surveillance is that the investigators were able to gather further details of the driver's trip from Colorado to Las Vegas, thanks to videos showing him from multiple angles taken by other vehicles then present at the charging stations. I'd be willing to place a healthy bet that few, if any, of the drivers of these electric vehicles were aware of the fact that they had a functioning onboard camera recording everything within camera sight. Of course all of this data was meaningless prior to the denouement: the explosion in Las Vegas that could have taken additional lives, if anyone else had been in the immediate vicinity of the Cybertruck.

This is yet another example of the opacity that results from combining surveillance with secrecy. Albert Fox Cahn, founder of the Surveillance Technology Oversight Project describes such hybrids (no pun intended) as "panopticons on wheels."

More than a decade ago, Apple released the location-sharing app Find My, which was originally intended as a way of keeping track of luggage. Soon users began relying on Find My as an anxiety-relieving technology that enabled them to locate their children. Still later, the app became the go-to technology for looking over the shoulder, metaphorically speaking, of romantic partners.

At the time of its introduction in 2013, 7 percent of iPhone owners used the app, with the percentage of users increasing in 2024 to 69 percent among members of Gen Z and 77 percent among millennials, compared to about 62 percent overall of adults in general.

Gen Z and millennials were the first generations to use technology as a means of tracking the location of as many of their peers as are willing to sign on for Find My. Some users track dozens or more people, a practice made all the easier once the app was automatically installed on iPhones and therefore easier to get parties to agree to be tracked.

As a result of this constant location mapping and sharing, a shift has taken place within the brain from a sense of autonomy to that of obsessive preoccupation with where "everybody is." Especially drawn to Find My programs are obsessive personalities, the chronically insecure "checkers" into friends' or lovers' habits. While in most cases nothing more is involved than an intrusive, "nosy" curiosity ("I see you were at the store yesterday. What did you buy?"), Find My is often used by stalkers and overly suspicious partners in romantic relationships.

Perhaps you are thinking that you would never consent to anything as intrusive as Find My. You might be surprised to learn, therefore, that your phone currently serves as a similar permission-less conduit for personal information.

According to the privacy practices of one of the nation's most circulated newspapers, data is used "to track you," including such data as financial information, contact information, search history, location, browsing history, and data usage, along with "data linked to you," which may be collected and linked to your identity, your purchase history, and finally, certain diagnostics that are not further explained on the privacy page of the *New York Times*.

The Find My app portends a more serious concern: How can we maintain any sense of autonomy when our location is immediately available for others to see?

In addition, the very nature of the Find My app leads to abuses and interpersonal duplicity. Consider this question sent in to the Q&A site Quora by a Find My user:

Question: "Can I fake my location, so my wife can't track my location?

Answer: "Yes, it's possible. A reliable option is to either remove the battery from the phone or wrap the phone in a sheet of aluminum foil, if your battery isn't removable. This will effectively put your phone in a Faraday cage, which will block ALL signals in or out."

In response to the prospect of always being "immediately available for others to see," as one Find My user complained, some users have opted for the creation of misleading or falsified location information. Their goal is to be able to identify and locate others while remaining geographically undetectable themselves. Here is the Quora suggestion on that: "If you need to demonstrate that you were some place like the gym, then you could try actually going to the gym and leaving your phone in a locker. If you have access to a building from work that others cannot get into, then you could just go and leave your phone on your desk."

Question: "I found out that my boss was tracking my location without my consent. I was so pissed at the moment that I deleted the tracking immediately. But I need to find a history."

Answer: "There is no history. If your boss was tracking you, then you are probably using a phone he gave you that is owned by the company you work for, and you must have allowed him access to your device. The only people that can see your phone's location, is if you gave them permission via the *Find My Friends* app, in which case they can always see your location."

The desire to mislead "trackers" can involve even more sophisticated evasions. In response to another Quora inquiry, a Find My subscriber was cautioned that "users with multiple devices may be sharing their location from a device they are not carrying."

(4) Smart Glasses

We've all heard about smartphones, smart kitchens, even entire smarthomes. The adjective "smart" implies the object is capable of receiving and transmitting information. But all of these smart instruments can potentially be employed to reveal and transmit more about us than many of us would feel comfortable revealing.

Smart glasses are particularly intriguing. The Ray-Ban Meta smart glasses look like any other pair of glasses, but they can stream pictures or video to Instagram or other preselected social media sites. Thanks to software developed by a third party, a backup computer can monitor the video stream and use AI to identify the face of the person photographed or videoed. These photos are then fed into public databases yielding the name, address, and phone number of the person currently being scanned by the smart glasses. Finally, all this information is funneled back to the smart glass wearer through a phone app.

Want to try a pair of Ray-Ban Meta smart glasses? In order to learn the practical details of how the glasses work, watch the Verge YouTube video featuring Victoria Song wearing the glasses and demonstrating and explaining to you how all the features work.

As you walk down the street, should you strap on your own smart glasses and become a surveillor, as well as a member of the increasing percentage of people being surveilled?

On New Year's Eve in 2025, Meta smart glasses—like many of the technological tools of the twenty-first century—were used for an unexpected, but not entirely surprising way. On that night at about 3:00 a.m., a forty-two-year-old Army veteran ran a truck into a crowd of revelers on New Orleans' Bourbon Street, killing at least fourteen people. After the driver was killed in a shoot-out with police, authorities found at the scene a pair of his Meta glasses.

Earlier in October and December 2024, the killer, later identified as an ISIS-inspired terrorist, had visited New Orleans and recorded footage of the area where, six hours prior to the New Year's Eve massacre, he planted explosives set to be activated by a detonator in the truck as it smashed into the crowd.

Thanks to the hands-free video recording made possible by his smart glasses, he was able to record during his October and December visits detailed footage of the area all without drawing public attention that may have accompanied the use of a cell phone to record in detail.

(5) "Maybe my neighbor really is a spy"

As a nation, we haven't experienced the outer limits of social surveillance. Yet. In China, surveillance was originally used as a crime-reduction strategy. In time and as a result of "surveillance creep" (the use of surveillance technology beyond its original purpose), surveillance became the centerpiece of a social credit rating system as an expansion far beyond traditional credit scoring systems. Financial and behavioral data is currently correlated with a "social credit" score. This is linked with facial recognition technology (FRT). You can think of FRT as similar to license plate recognition. If a police officer correctly notes a license plate, he can pull up reams of personal data concerning the person who owns the car.

In China, surveillance cameras have become such a feature of daily life that people are now accustomed to thinking that every human interaction taking place in a public space is probably the object of surveillance camera monitoring. Beijing alone has more than one million CCTV cameras. And the definition of public spaces has expanded to include anywhere outside of the confines of one's home.

China's leader Xi Chin Ping has touted national security as the country's top priority, ranking even higher than economic development. To reach the desired level of security, every Chinese citizen is urged to remain alert for threats as part of what is referred to as a "whole of society" mobilization.

"Right from the top it is perceived that the level of paranoia was insufficient among the general population," Andrew Chubb, a scholar of Chinese politics at Lancaster University in Great Britain told *New York Times* reporter Vivian Wang. In order to raise anxiety to sometimes a paranoiac level, citizens are instructed via a hotline about the ways everyday items may be employed for spying, i.e. pens with listening devices or cameras with the viewfinders set at right angles for capturing images off to the side of the object the camera holder is pointing it to.

The government has much to gain from instilling a deep and prevailing sense of anxiety in the population. But a delicate imbalance must be achieved: passive subservience to the government versus creating a level of anxiety and suspicion that constrains people-to-people contact and reduces their work efficiency. Certainly it's difficult to work at your greatest productivity if you suspect a coworker of observing or reporting on you.

"China's desire to put these scenarios at the top of people's minds and, make suspicion part of everyday life, is what sets it apart," according to Chubb.

When reading about surveillance in China, it's only too easy to assume people are aware, perhaps disapprovingly, but nonetheless aware of these surveillance measures. But anxiety can distort rational thinking, especially when it transforms into paranoia at a certain intensity: "Maybe my neighbor really is a spy. I've been feeling uncomfortable around him for a long time."

Over the years, I've treated many paranoid and paranoiac people and the progression is usually the same: It starts with an internal feeling of uneasiness, transitioning into a clinical state of anxiety, which when it reaches a certain intensity (different from one person to another) ushers in "a mental clarity" (a term frequently employed by the affected person to describe their experience). "I'm no longer anxious since I now have the explanation. I'm surrounded by people who cannot be trusted and may well be spies."

If you are over forty years of age, imagine how different you would now be, if from age twenty you were subjected to a surveillance model similar to the Chinese system. Every stroll in the park with your children; every meeting with friends at a bar or restaurant; every concert or lecture attended—all of these would have been recorded. In fact, in this imaginative exercise every person you interact with—even for the most trivial of purposes—forms part of your surveillance-compiled persona. If you associate with people with high social and credit rating, your ratings increase as well. But engage in a few interactions with people of less reputable ratings, and your rating may decrease in tandem. But nothing like this could happen here, right?

On September 12, 2024, Larry Ellison, Oracle cofounder and one of the half dozen richest men in the world, spoke to his company about a time in the near future when artificial systems would monitor citizens around the clock via an intensive network of cameras and drones. When asked about the benefits of such a system, Ellison—whether knowingly or unknowingly—channeled the early thinking of Jeremy Bentham in his response: "Citizens will be on their best behavior, because we are constantly recording and reporting everything that's going on," said Ellison.

One of Ellison's listeners pointed out that the mass universal public surveillance system Ellison suggested shares common features with George Orwell's now classic dystopian novel *1984*. But Ellison's model is vastly more efficient and creepy. AI systems instead of human observers would be in control, thus circumventing the biggest obstacle in any system requiring human participation: There are simply not enough people available to fill the roles of observers. If the camera records something thought worthy of intervention, someone has to note the offense and begin corrective measures. In contrast, AI, even when observing thousands of monitors at once, can home in on the person of interest. Nor are the agents of the law immune from Ellison's AI-grounded security system: "We are going to have supervision," he added. "Every police officer is going to be supervised at all times and if there is a problem, AI will report the problem to the appropriate person."

(6) Surveillance as an exercise of intimidation

"Surveillance has always been an exercise of intimidation," according to investigative journalist Ronan Farrow in the documentary film *Surveilled*, a 2024 HBO production directed by Matthew O'Neill and Perry Peltz. While looking much like his mother, actress Mia Farrow, Ronan has always been more drawn to real-life drama and during the better part of the past four years has focused on a specific proprietary spyware technology called Pegasus. Through one of its multiple third-party apps Pegasus can remotely activate and control a phone: Turn on the microphone, take pictures, record sound or video and capture any of the phone's data, photos, text, or location. All of these functions can be hacked without the phone user even suspecting it.

While its developer publicly claims that Pegasus was developed as an anti-terrorist effort, its use extends far beyond that. Spain, for instance, has employed Pegasus hacks to spy on suspected pro-Catalan independents, politicians, and separatist journalists.

Most worrying to Farrow are the clues he has been retrieving that suggest that the Department of Homeland Security has contracted with another spyware firm, Paragon, to break encrypted messages. Farrow has found that ICE (Immigration and Customs Enforcement) is also planning to use Pegasus spying technology as an aid in identifying and deporting illegal immigrants in large numbers. When he learned of this involvement by our government, Farrow referred to the Paragon purchase as an impending "digital panopticon," not just for the estimated 3.7 million people awaiting immigration hearings, but for the US population as a whole. Farrow's documentary *Surveilled* is available on HBO Max in the United States. For an inside glimpse of the most up-to-date surveillance systems and operations today, it is well worth watching.

On March 9, 2025, a French scientist flew from Paris to Houston in order to attend the fifty-sixth Lunar & Planetary Science Conference. Upon arrival in the United States, the scientist was asked to surrender his phone. Later he was told that he would be denied entry, because the search of the phone led to the discovery of critical comments concerning President Trump. He was informed that his "hateful and conspiratorial messages" had earned him both an entry denial and an FBI investigation. Hours later the scientist was told the FBI charges would not be pursued further but that he must return immediately to France.

In this instance, a private text message was forcibly revealed and punitive actions taken, even though the cellphone isn't usually thought of as "comprising social media."

No one was even aware of what he had said in criticism of the Trump administration.

Nonetheless the penalty was as swift and decisive as would have been expected for someone rioting in the streets.

Surveillance-facilitated penalties are already levied on groups for too candidly expressing opinions on social media. It started in earnest in October 2023 when trucks probed the campuses of Princeton, Columbia, and Harvard while exhibiting the names and photos of students and faculty who had allegedly signed statements of solidarity with Palestinians in Gaza. The organizers of these mobile vigilante forays on the campus called for the expulsion of students and firing of faculty. How were these alleged online statements discovered? By means of open-source intelligence researchers who had crafted into an art the monitoring of social media posts they deemed "anti-Semitic."

While anti-Semitic statements might seem at first difficult to define and identify, the definition was nothing short of crystal clear to those campus vigilantes: For them any criticism of Israel, however mild, was considered sufficient to meet the definition of anti-Semitism. Exposure of such allegedly anti-Semitic actions or statements was followed by punishments that were harsh and often life-changing. "Name, shame, and punish" is the mantra. Employers are contacted, "Do you knowingly employ anti-Semites?" Videos may be created; professional organizations brow-beaten into expelling the demonstrators; finally, in response to pressures comes the release of public statements by the school along the lines of, "We do not tolerate anti-Semitism in any form."

The largest and most influential organization presently engaged in this type of activity, informally referred to as doxing (publishing with malicious intent information on the internet about a person) is StopAntisemitism, which has over three hundred thousand followers and makes freely available the social media profiles and employment specifics of people deemed by that organization to be anti-Semitic. The overall effect was described by Laura Freedman, the president of the Foundation for Middle East Peace, in an interview with the *Washington Post*: "The intent here is not just to punish, but also to have a chilling effect. It's to send a message to people that . . . if you dare speak out of line when it comes to questions related to Israel you can and may face dramatic consequences-life changing consequences."

Let's skip over the political and cultural effects of the unbridled actions of such groups as StopAntisemitism. The psychological and neurological consequences are the key areas of interest here: the creation of anxiety in regard to expressing or even formulating opinions that lead to shaming and expulsions; an oppressive constricting sense of dread ("Have I said or written anything likely to bring attention to myself? Suppose somebody claims that I said something anti-Semitic, but I didn't?") The long-term harmful effects of having to subject oneself to such self-examinations include the creation of an internal observer, a parser of written words and statements, and eventually, even the formation or expression of certain thoughts all together. We've already reached a point in this country where certain statements simply cannot be made.

SURVEILLANCE

The figure at the beginning of this chapter of a surveillance network shows just some of the interconnections linking surveillance with anxiety, paranoia, violence, drug abuse, alcohol abuse, and criminality. The challenge for the twenty-first-century brain? Recognizing surveillance as a growing threat in our current society, something that, unfortunately, we must learn to live with in the short run.

Surveillance almost always leads to anxiety, except under circumstances when the surveilled remains unaware that they are under someone's observation.

(Continued . . .)

Initially, a mild sense of (1) uneasiness gives way to (2) suspicion, which, once it reaches a certain intensity, results in potentially incapacitating levels of (3) anxiety.

With the advent of (4) surveillance technology, surveillance became more intrusive and threatening.

(5) Surveillance *at work* leads to the conviction that one is always under observation. "Am I being watched now?" From here it's only too easy to enter the ominous realm of (6) paranoia.

Surveillance *at home* is often voluntarily but unknowingly self-determined thanks to the freely chosen connection of various appliances and devices to the internet, leading to the (7) "wired home" where everything from toasters to televisions provides potential insights into our personal habits and inclinations.

Not everyone can function optimally under states of anxiety. For those predisposed to acting out, (8) violence is propagated against individuals or groups identified as threatening.

In those with more optimally functioning frontal lobes, the response to anxiety may include (9) soothers such as alcohol, drugs, and tranquilizers which, if a person may be unable to acquire lawfully, may lead to reaching out to the shadowy denizens of the (10) drug underworld.

CHAPTER TEN

Anxiety

(1) An inescapable aspect of our lives

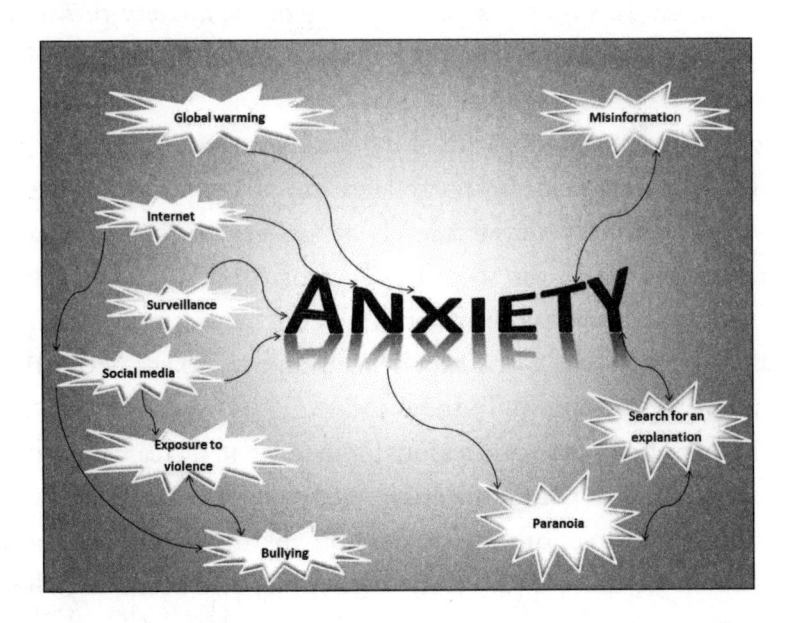

Life in the twenty-first century requires our brain to process a steady stream of anxiety-provoking events: protests turning violent, mass killings occurring at various locations around the world, infectious diseases that threaten to progress into modern day plagues, presidential-level assassination attempts (two attempts within two months aimed at then presidential candidate Donald Trump), tropical storms buffeting our nation's coasts, and computer hacking that puts at risk sensitive information regarding financial security, health matters, and privacy.

Anxiety has become such an inescapable part of our lives that even single words can arouse it. When I mention the word airplane, do you think about pleasant past trips and the prospect of equally pleasurable trips in the future? Or do you think about the much-publicized multiple incidents of near collisions between planes on the ground; turbulence episodes forcing planes to abort their flight; or outbreaks of fistfights and generalized mayhem in flight cabins, often leading to injuries to passengers and flight crews alike? This is only a partial listing of the anxiety-provoking events that are increasingly likely to be served up on any given day.

On page 189 is a diagram of the brain areas mediating anxiety, which has long been an integral feature of our lives. Americans reported strikingly elevated levels of anxiety in the 1990s compared to the 1950s, the so-called "Age of Anxiety." Among children, the situation was even worse. It's even been asserted that normal children in the 1980s experienced higher anxiety levels than adult *psychiatric patients* in the 1950s!

Experts point to several reasons for the current increase in anxiety.

Near-instant communication technology provides us with vivid video depictions of anxiety-provoking events occurring thousands of miles away, events that otherwise often bear little relevance to anything happening or likely to happen in our own lives. Media marketers take advantage of this by leveraging an important principle: If you want to get someone's attention, the best way of doing it is to arouse their anxiety. If you have any doubts about this, just watch the nightly newscasts or read any of the major national morning papers. Everything seems so dire, as reported in these outlets, based on the well-known psychological principle that most of us pay more attention to those people who speak to us of the terrible things that may happen or are happening, than we do to people who assure us that everything is fine. This predilection is based on the brain's need to respond more quickly to potentially harmful situations than to innocent ones. False positives (responding unnecessarily) may be inconvenient, but false negatives (not responding to a real threat) may be deadly.

As a result of communication technology, we are now exposed to information about innumerable nerve-racking calamities. To make matters worse, anxiety tends to be cumulative: If we become more anxious about something today, then our anxiety will likely resurface whenever we encounter that same event, similar event, or situation in the future. And since each day provides any number of anxiety triggers, anxiety arousal increases at a steady rate over the years.

(2) The chronic fallback emotional state

What's the difference between fear and anxiety? Informally, fear is often thought to require a person or an object: the bear you weren't expecting to encounter when you opened the garage door of your weekend getaway in Maine. Anxiety in contrast may or may not be associated with a specific object, topic, or event.

Although this traditional distinction remains generally applicable today, it's an oversimplification and you don't have to be a psychologist or psychiatrist to recognize this. Sometimes anxiety and fear intermesh with each other so intimately that it is difficult to distinguish one from the other.

For example, between the spring of 2020 and the present most of us at one point or another were fearful of contracting

COVID-19. This was understandable based on the progression of the illness from a localized epidemic starting in China to a worldwide pandemic. But if we dwelled too much on the possibility of contracting COVID, we segued into a chronic anxious state, which took on a life of its own.

During the height of the COVID pandemic (2020–2022), the incidence of insomnia tripled among young adults, along with a vast increase in mental health crises and a faltering in both workers' and students' capacity to control and appease distressful emotions like anxiety. As a result, anxiety-associated conditions skyrocketed. None of this was the result of contracting COVID, but rather the result of the anxiety associated with the *possibility* of contracting it.

A similar situation exists now in response to global warming. At some point if you read enough about it, you'll experience the kind of anxiety that arises when an elevator door takes longer to open than you expected: "Am I trapped in here?" flits across your mind. Or the anxiety may be prolonged, lasting minutes rather than seconds.

Anxiety has become the chronic fallback emotional state for many of us in response to events and/or the interpretations we put on them. In response to the election of the new president, one of my psychiatric colleagues, the day after Election Day, treated two of his patients for acute anxiety attacks.

(3) Brain and anxiety

What is going on in the brain during anxiety? Reduced to its essentials, beneath the overarching cerebral hemispheres lies a network of brain structures, the limbic system, that, along with certain hemispheric areas such as the cingulate gyrus, provide the underpinnings for emotions. While our *experience* of emotions arises thanks to *activation* within this network, the *expression* of emotions depends on the contributions from the cerebral hemispheres, specifically the frontal cerebral cortex. While we retain only limited control over the *experience* of our emotions, we remain, in most instances, largely in control of our resulting emotional *expressions*. A moment's reflection on everyday experiences demonstrates the interplay of limbic and select cerebral hemisphere areas.

If during a staff meeting someone makes a snide deprecatory remark toward you, how do you respond? Under such circumstance you could experience during the first few seconds the impulse (it is an impulse: an emotional one) to speak sharply with something equally derogatory directed to the speaker. But if your brain is functioning normally you will also experience a "reality check": You'll rapidly determine that you have more to gain by resisting the impulse to counterattack and, instead, stick to the issue under discussion.

It's the cerebral hemispheres, specifically the frontal cortex, that provides the power to restrain your initial impulsive responses. Those with abnormalities of the frontal cortex, such as frontal lobe dementia, aren't able to monitor their emotional responses and lash out in retaliation.

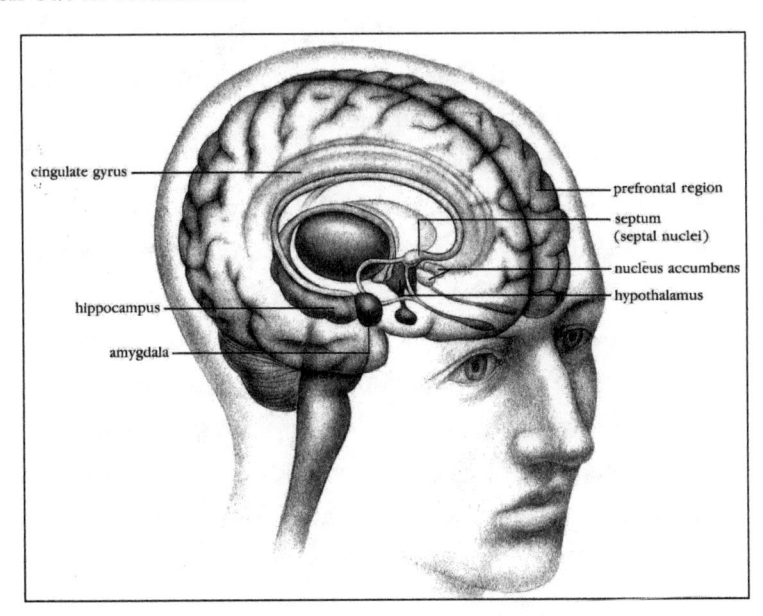

This diagram illustrates the anatomy of the frontal lobe/limbic system brain areas. Under optimal conditions, both brain areas are in perfect balance. Fatigue, lack of sleep, alcohol, limbic system seizures, and lingering resentment can tip the balance toward excessive inappropriate expressions of emotion. Mindfulness and looking at the "big picture" ("An emotional outburst by me is going to get me nowhere with this office team") signals that the frontal lobe is now in play.

Consider just some of the twenty-first-century contributors to redefining anxiety from a pathological condition, experienced principally by people suffering from one form or another of mental dysfunction, to a component of a "normal personality"—if such a standard as "normal" now even exists.

Reading over the following chronology of dates and events, consider for yourself what effects these might reasonably exert in creating anxiety. As one person put it when his anxiety had gotten out of control—"I'm now always feeling I'm unsafe and helpless against something I can't put into words."

- **2001:** The September 11 Al Qaeda suicide attacks on New York City and the Pentagon. The date is so engrained in our national identity that merely typing September 11 into any search engine brings up September 11, 2001.
- **2007–2008:** Losses in mortgage-related financial assets strained global markets, leading to a recession in the United States by December 2007. Home prices fell by over a fifth on average nationwide from 2007 through 2011. Although the recession officially ended in June 2009, economic weakness and uncertainty persist to this day with potential homebuyers contending with fluctuating interest rates and rapidly increasing prices.

- **2009:** One of the first of many subsequent climate warnings to capture widespread attention came in a report titled "A National Infrastructure for The Twenty-First Century" published in the United Kingdom by the Council For Science & Technology. The document so well reflects today's world that when reading it you could be forgiven for thinking it was written yesterday rather than sixteen years ago: "The effects of climate change are predicted to cause higher summer and winter temperatures, sea level rises, a rising intensity of storms, forest fires, draughts, increased flooding, heatwaves . . . Climate change is potentially the greatest challenge to global stability and security."
- **2011:** For fifty-one days from September 17 until November 15, 2011, Occupy Wall Street protestors opposed social and economic inequality. The movement served as a stimulus for protest movements to come.
- **2016/2021:** The election of Donald Trump. Whatever your political affiliations, you will probably agree that the years from 2017 to 2020 were marked by turbulence, concluding with the storming of the Capitol on January 6, 2021.
- **2020–2022:** The COVID pandemic. An event whose origins remain unknown: a microbe escaped from a laboratory, or a crossover of a microbe from an animal (a bat is considered the primary contender) to a human? In the absence of any agreed-upon explanation, controversy and misinformation continue to abound even now. The pandemic also proved to be a significant instigator for the interweaving of a vast number of conspiracy theories.
- **2021:** After the in-custody murder of George Floyd on May 25, 2021, protests, both peaceful and violent, broke out in over 140 cities, triggering National Guard activation

in more than twenty states. Shootings, violence, and vandalism appeared on national television with the same scenes repeated over and over in an endless loop. Platoons of marchers roamed neighborhoods, prompting home and business owners to place outside their homes and businesses Black Lives Matter posters in response to a perceived threat.

(4) "Helpless against something I can't put into words"

As a result of these occurrences, along with many others in the first quarter of the twenty-first century, our entire culture is experiencing a brain-facilitated sense of fragility, threat, and peril. As examples, consider the following themes as reported innumerable times on the evening news:

- Global warming is held responsible for storms and tornadoes sweeping across the country provoking historical deluges and floods, depicted in vivid video images.
- Violent shootings such as the one that occurred in Ohio on March 25, 2024, when an Uber driver was shot and killed by an eighty-one-year-old man who had been previously instructed on the telephone by scammers to gather together a large sum of money, put it in a package, and give the package to the Uber driver, who was on her way to pick it up. The driver in turn was called by the same scammers and instructed to pick up the package. She was told nothing about the contents of the package. The news segment features a video loop showing a terrified black woman backing up while pleading for her life as an elderly white man with a pistol advances on her at a steady pace. Several seconds later the sound of gunfire can be heard off camera.

- In other segments, random stabbings are shown with grisly examples drawn from cities extending from Los Angeles to New York City.
- Global wars are taking place in Ukraine and the Gaza Strip, where videos show gruesome images of small Palestinian children and mothers being shot by members of the Israel Defense Force.

Does the problem arise from these extreme events or from the effects of these depictions on our brains, leading to disturbances in our own mental health? In other words, is it the events themselves as we might read about them in a newspaper or magazine, or is it attributable to our repetitive exposure to vivid high-quality images of natural disasters and horrific scenes of human violence?

(5) "If it bleeds, it leads"

With the advent of high-resolution cellphone cameras, macabre violence-laced images started to be recorded on the spot and instantly conveyed over the internet to television networks for national and international dissemination based on the most fundamental principle of video journalism: "If it bleeds, it leads."

How often on the nightly news have you heard these words spoken by the moderator in a tone of faux concern: "We want to warn you the images you are about to see may be disturbing." What the anchor doesn't tell you is that the parameters of what can prove "disturbing" are shifting like fish caught in an undertow. What's determined to be "disturbing" is based principally on what has been broadcast before without eliciting widespread objection. While the viewer may ruminate about what he has seen and may even lose some sleep over it at some point—typically most rapidly among those exposed to the greatest number of "disturbing images"—indifference and apathy eventually intervene. Even the most seasoned consumer of violent imagery runs the risk of developing what is known as secondary trauma, according to Roxane Cohen Silver, a professor of psychological science at the University of California, Irvine. Professor Silver has made a career of studying the mass effects of violence among those who watch televised coverage of violence repetitively.

Three major findings have emerged from Silver's research. First, the greater number of hours a person spends immersed in video presentations of violent scenes, the more likely they are to experience acute stress and post-traumatic stress symptoms. Not only mental health symptoms (anxiety, panic attacks, or disturbed sleep sometimes accompanied by playbacks of scenes from the televised screen feed), but physical symptoms can also arise accompanying the exposure.

Silver's second finding is unexpected and surprising for most people upon first hearing it: Individuals choosing to watch and rewatch scenes of a horrific crime may actually experience more distressful symptoms than people actually present on-site when the crime was committed. The Boston marathon bombing in 2013 resulted in greater harm among television "junkies" than among people who had been on the scene that day. Alternately, repeated exposure leads other people to indifference and apathy.

Finally, an addictive quality seems to be present: The more distressed and worried one may be after reviewing television coverage of a horrific experience, the more likely a person will seek out media exposure—exacerbating the symptoms even more to the point of post-traumatic stress disorder (PTSD). "Doomscrolling" is the technical term for this semi-addictive attraction to catastrophes.

(6) A child hit by a car

In discussions about PTSD, one question has never been resolved: Does PTSD by proxy exist? In other words, can a person suffer serious mental health consequences following exposure to graphic depictions of horrific images of bodily trauma, suffering, and battlefield casualties? And if so, what are the implications?

These questions take on a special importance at a time of heightened press coverage of worldwide wars and uprisings, especially in Ukraine and Gaza. In these battle zones, we are directly involved in providing the military weaponry that contributes to the horrific images that are launched like grenades into our living rooms.

Let's identify an important misconception. According to the current understanding of PTSD, the affected person must actually experience the potentially fatal event. A picture or someone else's description of the event is insufficient. But does that really make sense?

Imagine a man arriving at home from work one early evening. As he gets out of his car, he notices that the neighborhood seems uncharacteristically quiet. He feels vaguely uneasy, but can't come up with any reason for his disquiet. His wife or son who usually come out to meet him upon his arrival are nowhere in sight.

At this point he sees his neighbor's front door open. He watches uneasily as the neighbor approaches him—grim faced. The neighbor relates in graphic detail how the man's five-year-old son had been run over by a car only a half hour earlier. The neighbor relates that the man's wife went off with the injured child in the ambulance. There had been no time to call the father on his cellphone. Worst of all, the neighbor awkwardly blurts out that she had heard one of the members of the EMS team say, "The little guy might not make it."

Now if that boy dies, can the father experience the same acute stress response (the original prodromal symptoms leading to PTSD) as the mother, who had actually witnessed their son running into the street and being dragged under the wheels of a car? I think he could. The same risk would hold if he later viewed a video taken as the accident occurred.

Which brings us back to the televised images from Ukraine and Gaza. Not everyone agrees that witnessing carnage on a TV screen instead of "in real life" would be sufficient to result in PTSD. As journalist Hannah Fern wrote in an article in *The Independent*, "You might be tempted to look away from the photo of Ukraine—don't." When the image becomes difficult to look at, she wrote, "resist that temptation, shares those images, talk about what you see."

Although her advice may be appropriate for professionals trained to carry on in the face of sometimes hideous depictions of trauma, other people may be disturbed to the point of extreme anxiety or even PTSD in response to these explicit depictions. For them, looking away, or turning off their television, may be the best courses of action. And things are only likely to get worse.

Thanks to continuing advances in the resolution of images and widespread, almost universal, availability of the internet, we can see destruction, suffering, and casualties with a clarity never before encountered. In previous military engagements or wars in the twentieth century, such as Vietnam, less technologically sophisticated images provided disturbing experiences for the viewer. For instance, most historians agree that the televised images of the fighting in Vietnam served as a spur to the success of the anti-war movement, which eventually led to the ending of the war. But bear in mind, this occurred prior to today's superpowered images and, of course, well before the internet.

One wonders what would have been the national and international mental health consequences if large proportions of the public had been glued to live streaming videos of Auschwitz or

Treblinka. Even the grainy, low-definition photographs from those death camps available to us today induce an unspeakable horror.

(7) Traumatizing a population

In some instances, horrific war footage is being deliberately aimed at select populations. For example, extremely graphic images of death and destruction were broadcast early in the Ukraine war into Russia. As *Washington Post* reporter Drew Harwell described: "A constant stream of extremely graphic images showcasing the horrors of war and inviting Russians to examine them to determine whether the images feature a missing loved one." In many of the images, he added that "Soldiers' corpses can be seen burned, ripped apart, wrangled in wreckage, or abandoned in snow; some, their faces are featured in bloody close-ups, frozen in pain."

If horrific images can be used to deliberately traumatize a population, would accidental or casual exposure, say by channel surfing, bring about the same result? What about the effect of a single photograph? The front page of the June 10, 2024, *Washington Post* featured one of the most distressing pictures I've ever encountered in a newspaper. It depicted a young Palestinian girl climbing over debris in Gaza's Nuseirat Refugee Camp on Sunday, June 9, 2024, the day after four Israeli hostages were rescued at the cost of 270 Palestinian lives.

Given my experience in diagnosing and treating PTSD, I fear that a significant portion of people exposed to such images from both the Israeli-Palestinian war and the Russian-Ukrainian war are at risk of suffering some form of severe mental distress. Even as a trained neurologist and psychiatrist, I found these images disturbing. We can't yet conclude for certain whether images of other people's suffering, mutilation, and death can portend great consequences for our own mental health, our sensibilities, and our shared sense of humanity. But at the current rate of exposure, we'll soon have our answer.

(8) Can nuclear war be avoided? For how long?

When we speak of anxiety and its causes, one subject dominates all others: the fear of nuclear war. The tensions surrounding the missile threats from Cuba in the 1960s, or the prospect of nuclear war just prior to the tearing down of the Berlin Wall on November 9, 1989, were less risky than the present situation for one reason: good leadership. Compare Kennedy and Khrushchev, Reagan and Gorbachev, to Trump and Putin, who, at the time of this writing, are like two dancers with grenades in their hands pirouetting closer and closer together in a mad dance. Only they aren't holding grenades, but nuclear bombs.

A US versus Russia nuclear war is universally recognized as likely to lead to Armageddon. Less recognized are the effects of a nuclear exchange between any of the other seven nuclear-armed combatants. It would alter life on our planet beyond anything previously experience or even imagined.

Take the India versus Pakistan standoff in May 2025. Fortunately, hostilities ceased after an alleged US-brokered ceasefire on May 11. It's useful to imagine what would have been the consequence for the rest of the world if India and Pakistan had launched their nuclear weapons. But we don't just have to *imagine* the consequences.

A scientific study in 2019 assessed the potential consequences of a nuclear war between India and Pakistan. It found, according to W. J. Hennigan, a national security authority and a reporter for the *New York Times*, who wrote in his May 15, 2025, article "Nuclear War Avoided, Again. But Next Time?" that "the detonations would create millions of tons of soot. Clouds of debris would block out the sun and lower global temperatures to bring about worldwide famine. Millions of people would die and billions could be affected."

The study from 2019 that Hennigan refers to contains the following information: If the bombs used were one half of the power of the bomb dropped on Hiroshima, there would still be a massive smoke emission that would go into the upper atmosphere and remain there for perhaps decades. Fifty to one hundred and twenty-five million people would die, many large cities would be destroyed or rendered uninhabitable, and millions would be injured and in need of care. In addition power, transportation, and financial infrastructures would be reduced to ruins. Bear in mind that currently nine nations are known to possess atomic weapons, with a global nuclear stockpile estimated at around thirteen thousand weapons.

(9) LA fires and Hiroshima

For good or ill, we are now capable of answering the most anxiety-provoking question of all: What might the world look like after the explosion of one or more of these thirteen thousand accumulated nuclear weapons? In response, many of us would turn to photographs of Hiroshima. But there isn't any need to do that. The evidence is right here, right now.

For twenty-four days from January 7 to January 31, 2025, we witnessed in miniature a close approximation of what our world, specifically the United States, might look like after a nuclear explosion. "The fires of Hiroshima and Los Angeles: Apocalypse redux," an essay in the *Bulletin of the Atomic Scientists*, links the two events (Hiroshima and the LA fires) to provide a deeply unsettling possibility that we are forced to wrap our minds (our brains) around.

For one thing, nuclear experts agree that massive fires would occur immediately following a nuclear explosion in any major city, according to the authors of the *Bulletin of the Atomic Scientists* article, Richard Turco and Owen Brian Toon. What's more, if you compare the Los Angeles fires with an atomic explosion, "the total area is comparable to that of the primary destructive zone of a strategic nuclear weapon." The key difference between the two events pertains to the accompanying casualties. The total fatality of the Los Angeles wildfires, twenty-nine people, "pales in contrast to the hundreds of thousands of deaths expected in a nuclear attack . . . and it is not unlikely that up to a million individuals would eventually succumb."

If you watch the nightly news with any consistency, you have no doubt viewed hours and hours of video recordings of the Los Angeles conflagration. It's likely too that at various points in your life you've looked at photos of the aftermath of the Hiroshima explosion. As a reminder of that horrific day, enter the name Yoshito Matsushige into your favorite search engine. Matsushige was a Hiroshima survivor and a Japanese military photographer, who provided the only known photos of the aftermath of the August 6, 1945, nuclear bombing of Hiroshima. Several of them look very similar to the photos taken after the Los Angeles fires. One photo taken by Matsushige of Hiroshima two months later in October 1945, titled "A camphor tree fallen by atomic bomb's blast," could easily be mistaken for a photograph taken after the Los Angeles fires: a huge tree uprooted and lying among the ruins of a browned, burned-out structure that looks for all the world like a house.

Did you link these separate events (Hiroshima and the LA fires) in your mind, so that seeing one reminded you of the other? I expect you didn't in any conscious way; I know that I didn't. But I'm reasonably confident that the authors Richard Turco and

Owen Brian Toon are not the only people who conjured this compelling, anxiety-provoking analogy. Such comparisons often percolate in our subconscious experience. And when they arise into conscious awareness, they prove to be the most fearsome and anxiety provoking of the five existential threats: pandemics, climate change, habitat and diversity loss, unregulated AI, and, finally, nuclear weapons. Each interacts with the others and can't be considered in isolation.

Now that Turco and Toon have pointed out to us the similarity between the fires of Hiroshima and the increasing number of forest fires—not just in the United States but worldwide—I doubt you will ever be able to look at the destruction caused by a forest fire and not conjure in your imagination, however briefly, the specter of a nuclear explosion.

Most worrisome of all, portents for the detonation of a twenty-first-century nuclear weapon are all around us: the recent push for atomic energy as a power solution for the increasingly avaricious demands of advanced AI, the return of uranium mining in the American Southwest, and the increase in "loose talk," principally by Russia, reminding its perceived enemies about the potential deployment of nuclear weapons. Even more anxiety provoking, all of these events are taking place in the current setting of rapidly spreading isolationism, populism, misinformation, disinformation, and short-term thinking

The last chapter of this book will take up several of the options available to prevent the persistence, even worsening, of the twenty-first-century challenges now facing the brain.

CHAPTER ELEVEN

New Ways of Thinking

(1) Searching for hidden links among seemingly unrelated topics

During the remainder of the twenty-first century, we will all have to train our brains to new ways of thinking in order to solve the problems we are facing. The most basic operation: searching for hidden links among seemingly unrelated topics, often unexpected ones. As a consequence, new concepts and responses are required to discover, describe, and deal with them.

I observed firsthand the importance of hidden links during a course on the brain that I taught at the FBI Academy in Quantico, Virginia.

Before and after lectures the agents discussed with me several serial killers and how their apprehension depended on linking seemingly trivial threads of information (tire marks, empty soda cans, cigarette butts of a particular brand, etc.). While these findings didn't themselves identify the killer, they did provide an entry point to be associated along with other clues pointing to the specific serial killer.

As a neurologist and neuropsychiatrist, my most avid interest was in the psychological and behavioral profiles the agents had assembled prior to the killer's identification.

Two points struck me at the time. Each of the agents was specialized in a particular area. Second, and related to the first, the identification of the killer could not have been accomplished by any one agent working alone.

Sometimes the locale where the killer operated was limited, only a matter of miles, as with serial killer Dayton Leroy Rogers, whose murder spree was confined to Portland, Oregon, and its immediate surroundings. (I mention Rogers because I became intimately familiar with the details of the killings when serving as an expert witness in his trial.)

In other instances, the killers spread their grisly work over widely scattered locations throughout the country. Among the killers who committed separate murders nationwide, the true nature of the crimes only became clear after multiple experts in different forensic specialties cooperated to link the seemingly unrelated killings to one person.

A similar solution to the challenges discussed in this book will only come about when everyone agrees on a shared purpose best served by cooperation. But as I readily recognize, the comparison between identifying a serial killer and solving something like global warming isn't entirely valid.

Law enforcement officers, specifically in the cases I encountered at Quantico, shared a common goal: discovering the killer. In contrast, nothing approaching unanimity exists in our societal approach to global warming. Instead, we observe vast variation in response, ranging from complete denial to a climate-control-lite approach that grudgingly acknowledges the problem but lacks the commitment and willingness to take the necessary steps to control it. Political points of view also differ. During the Biden administration, nascent positive measures were started, which, at least in the early days of the administration of Donald Trump, have been largely reversed. In the meantime, global warming continues to simmer.

Nothing less than a new way of thinking will be required to solve global warming and the other twenty-first-century problems discussed in this book. What's needed along with reductions in the burning of fossil fuels are other approaches, aimed at controlling global warming. What specifically will this entail?

Up until now each of the challenges posed to the twenty-first-century brain has been considered mostly in isolation. Experts on the physical challenges (global warming, infectious diseases, the internet and AI) may vaguely appreciate the importance of cultural drivers like surveillance and stress. But they are not trained in psychological and social sciences and, truth be told, they often have little interest in them, preferring instead to concentrate on the strictly scientific disciplines they've been trained in.

I'm reminded of the Hindu story of a group of blind men trying to come up with an appreciation of an elephant's "form" by a piecemeal approach: one person palpating the trunk, another the tusks, others running their hands over the dermis at various bodily points? Not an impossible approach, but a very difficult one if you want to appreciate the animal as a whole. A similar situation exists when trying to address any of the twenty-first-century challenges discussed earlier in this book without considering the contributions of many others.

Responding to one challenge at a time is a nonstarter. As I hope I established throughout this book, the challenges are so intertwined with each other that selecting one and working toward solving it in isolation without considering the contributions of many others makes the solution even more complicated than the original problem.

(2) The need for a "mental upgrade"

One of the paths we can follow was first articulated in 2018 by the innovative and brilliant philosopher Timothy Morton, who, in 2013, coined the term *hyperobject*. "I invented a word to describe all kinds of things you can study and think and compute, but that are not so easy to see directly," wrote Morton.

According to Morton, this new word applies especially to global warming. Thinking about global warming requires a kind of "mental upgrade," according to Morton, in order to cope with "something that is so big and so powerful that until now we have had no real word for it."

Global warming is a *hyperobject* that can't be seen or touched directly, but "it's really widespread and really long-lasting (100,000 years), but it's also super high-dimensional."

As a key feature, hyperobjects don't involve single objects, but the *relations* between several, even many, objects. For example, global warming is currently understood as resulting from interactions involving sun, fossil fuels, carbon dioxide levels, ocean warming, emission levels of CO_2 and methane, along with probably any number of other variables. Measurements and models of these interactions among these contributors constitute the hyperobject we refer to as global warming.

Since a hyperobject is a difficult concept to grasp, here is my analogy (not Morton's) that may clarify things.

Imagine that you and I are competing against each other in a timed memorization contest involving the following list of words:

> javelin
> cereal
> willow tree
> mustache
> pocket watch

tightrope
Irish whiskey
television
yo-yo
sentry

Stare at the words for two minutes and then see if you can recite them all. Unless you've read my previous book, *The Complete Guide to Memory* or a similar book, or were born with a highly proficient memory, you likely won't be able to do it. But I had no problem memorizing that list (made up, incidentally, by someone other than myself) in two minutes. If you too were able to do that, add five more words to the list. I have no difficulty doing that either. How well did you do?

If you didn't do as well, how do you explain this discrepancy between us in memory facility? You might correctly answer that I'm probably using some kind of "system." But that insight doesn't provide you with the method that I used nor does it lead to any clues how you may improve your performance.

In order to narrow the possibilities a bit, I can tell you that the memorization didn't depend on any similarities or distinctions among the words: It didn't rely on the number of the words' syllables or their arrangement (I can memorize them and repeat them back in any order requested). My seemingly mysterious method can best be thought of as a *tension*, a form of mental kinetic energy—a kind of force field—uniting all of the words (think of fine threads or tiny wires). Knowing what that kinetic energy involves provides the answer to how I memorized them in such short order. So here is my method:

As a means of quickly memorizing them, I mentally enacted a two-act play. In Act One, a *mustachioed sentry* is starting his day while sitting outside at a table beside a *willow tree* watching *TV* while eating his *breakfast of cereal* and *Irish whiskey*. (As a sentry he is a rough-and-tumble guy who prefers Irish whiskey to orange juice.)

When the sentry has finished his breakfast, Act Two begins with the sentry exercising by walking a *tightrope* while holding a *javelin* in one hand while yo-yoing with his prized *yo-yo* in the other. Act Two ends with the sentry getting off the tightrope and pulling out his gold *pocket watch* to check that he will not be late for duty.

Once you've read that mini description and mentally visualized that two-act play, you can clearly see all of the words in your mind the way they exist in mine. Recitation of the ten items will then be easy. You can even recite them backward or forward or in any configuration you choose.

In order to memorize the words, you only had to unite them in a narrative consisting of pictures rather than words. Think of that narrative uniting the different words as the hyperobject: It can't be seen or appreciated by any of the senses.

In this experiment, the unseen and unheard mental mini-play taking place in my brain includes all the words on the list and brings them to life in an easily remembered mini-drama. Think of the forces influencing the interaction of character and situations as corresponding to the hyperobject.

As with the mini-drama, we can only observe hyperobjects in terms of a constellation of mutual influences, but we never see the actual hyperobject itself but only its components, because at the most basic level, a hyperobject is not a *thing* but a phenomenon. In the example, the relationships of the objects in the mini-drama are held together by the force of my imagination uniting the words into the elements of a two-act play.

Examples of hyperobjects include inflation or nuclear exposure levels. While results of each can be perceived (higher prices in the supermarket, higher radiation detected by special instruments), the concepts themselves remain just that, concepts. Grasping the nature of a hyperobject requires a level of abstraction that so far has proven challenging to our twenty-first-century brain.

"One of the main characteristics of hyperobjects is that we only ever perceive their imprints on other things," according to James Bridle, author of *New Dark Age*. "Hyperobjects can only be appreciated at the network level."

Since hyperobjects cannot be touched or otherwise perceived through our senses, it's a tremendous challenge for our twenty-first-century brain to conceptualize or even think about them at all. Yet in order to survive the current and future threats and stresses facing us, we are going to have to learn to discover the invisible but logically discoverable connections forming the hyperobject. We are talking here about a new way of thinking based on our new understanding of the brain as a connectome. Applying this to climate change, Bridle stated, "Arguments about the existence of climate change are really arguments about what we can think."

As an example of the conceptual challenges we currently face, many people experience difficulties distinguishing weather from climate trends. You no doubt have had someone say to you something along the following lines, "Last winter we had several snow days and more than a few really cold days. So how can anybody say that the whole world is warming?"

This question highlights a vulnerability in the brain's innate power to distinguish between weather and climate.

While we pay careful attention to local weather events and plan accordingly, equally probable climate trends extending over decades are often ignored. While we've all experienced a day or two of thunderstorms, prolonged weeklong downpours are rare but are becoming more common. But even this way of thinking over the span of decades is dangerously outdated.

Experts from around the world assure us that future floods will become increasingly frequent and intense. So why is that information so difficult for us to mentally assimilate and plan for future occurrences? Because as James Bridle suggested: "We have to stand outside both our perception and our measurements. Because hyperobjects are so close and yet so hard to see they defy our ability to describe them rationally in any traditional sense. Taking this a bit further, hyperobjects can only be appreciated at the network level, manifested through vast distributed systems of sensors handling huge amounts of data, and computation."

Another way of conceptualizing hyperobjects is to consider them similar to what novelist Hari Kunzru calls AST novels—apocalyptic systems thrillers. "These books invite us to take an elevated panoramic view of *events that extend too far in space and time to be grasped by a single narrative consciousness*," wrote Kunzru in an essay in the *New York Times Book Review*. "Conflict, climate change, pandemics and natural disasters offer ways to contemplate our interconnections and interdependence." [ital. mine]

A simplified way of illustrating this interconnection and interdependence is by linkage diagrams such as those on page 41. The central concept here involves the effect of *weather events*, whether they be heat waves, hurricanes, or forest fires. Any of these catastrophes can result in loss of life and destruction of property, each of which is linked to additional negative outcomes. Those who are killed or maimed can no longer be employed or pay taxes. In addition, the destruction of property leads to huge governmental and financial outlays (funds are still being directed toward the losses resulting from Hurricane Katrina twenty years ago) that further increased our already outsized national debt. Also weather events are creating a new population subclass: homelessness among members of the middle class.

Stare at the diagrams on page 41 for a few minutes. All of these interconnections and interactions can only become graspable when they are laid out in outline form such as a linkage diagram. Reading verbal descriptions such as in the previous paragraph results in grasping the components as separate entities. But they are not separate entities and can only be understood in terms of their interactions as in the diagrams.

One of my suggestions, when you are thinking about the challenges faced by the twenty-first-century brain, is to construct your own linkage diagrams. The goal is to remain aware that nothing ever occurs without reference to an undetermined number of other things. The diagram on page 41 brings to the foreground only some of the multiple effects of weather events. Of course there are many others, but my point is that we should think of these seemingly separate events as components of vast interlinked and interconnected networks corresponding to the hyperobject. Similar vast interconnections exist within the brain: the connectome.

Please read that last sentence again because it describes not only hyperobjects, but the brain that intuits them. In fact, you can think of the human brain itself as a hyperobject arranged in the form of a connectome. It's been one theme of this book that understanding the seminal challenges of the twenty-first century (global warming, AI, etc.) will require a new understanding of the brain as both (1) a connectome; and (2) a hyperobject.

(3) Occam was wrong

Many problems are best solved by limiting the number of variables included in the solution. This principle was first formulated in the fourteenth century by philosopher William of Ockham: "Entia non sunt multiplicanda praeter necessitate," which translates to "Entities must not be multiplied beyond necessity." Violations of

Occam's razor involve invoking more components in the solution than are required. But an equally egregious violation consists of arbitrarily picking only one of many possible components and concentrating on just that one to the exclusion of all the others. Environmental pollution is a case in point. But environmental toxins don't exist in isolation from each other. In almost all cases they are additive.

Researchers from Johns Hopkins University published a paper in March 2025 measuring the cumulative effects on health of multiple toxic components. In doing so, researchers measured thirty-two hazardous air pollutants in communities in southeastern Pennsylvania located near petrochemical facilities. The pollutants included vinyl chloride from formaldehyde and benzene. By developing profiles of the pollution concentration in the air at specific times, they could determine what toxins people were breathing at any given time. To everyone's surprise, thirty-two hazardous air pollutants occurred concurrently, each adding to the cumulative effects on the functioning brain.

"When we regulate chemicals, we pretend that we are only exposed to one chemical at a time," Keeve Nachman, the study's senior author, told science writer Amudalat Haasa. "If we have each chemical and we only think about the most sensitive effect, we ignore the fact that it could potentially cause all of these other effects to different parts of the body."

That is also true when it comes to understanding what effects can be caused by the various challenges discussed in this book. We have to think polyphonically. Environmental pollution and the other challenges are not solo performances. All of them are occurring concurrently, while inducing cumulative effects on the functioning brain.

What's more, our twenty-first-century demons tend to reinforce each other. Crowding due to a population explosion, increases in global temperature, and COVID-19, for instance, served as a "witches' brew" in 2020.

Six months into the COVID lockdowns at the beginning of a torrid summer on Monday, May 25, 2020, at 8 p.m., George Floyd exited Cup Foods in Minneapolis, Minnesota. A store clerk called the police and asserted that Floyd had paid for cigarettes with a phony twenty-dollar bill. Floyd was wrestled to the ground followed by Officer Derek Chauvin pressing his knee into Floyd's neck for nine minutes and twenty-nine seconds. Floyd gasped that he couldn't breathe. Nevertheless the knee application was continued for two minutes and fifty-three seconds after Floyd had ceased to put up a struggle. At that point, Floyd was dead.

From May 26 to June 28, 2020, between fifteen million and twenty-six million people participated in demonstrations organized and facilitated by Black Lives Matter (BLM). Almost half of the cities in the United States experienced demonstrations in which violence erupted. Was the co-occurrence of COVID and protesting accompanied by the eruption of violence just coincidental? Perhaps. But racism in all of its naked ugliness had been present, to a greater or lesser extent, in the United States for over two hundred years. So, why did this eruption occur when it did? Since the brains of the participants in the riots were already coping with the stress of COVID lockdowns leading to a simmering anxiety and anger, you have the elements needed for a riot. When the brain is overcome with unmanageable anxiety, everything becomes threatening. Under such conditions, the pent-up fear and anger only too easily segue into violence.

As with the George Floyd episode, it is necessary to take into consideration not only global warming, which we discussed in detail earlier, but also the political, social, racial, and environmental tensions currently dividing our society to understand things like the increasing incidence of periodic outbreaks of violence.

(4) Volatility, uncertainty, complexity, ambiguity

In the 1980s, the US Army War College described what they termed the VUCA approach to the uncertainties of the modern battlefield. The goal was understanding and responding to a battle situation that was "volatile, uncertain, complex, and ambiguous." Little did those innovators know that by the early 2020s the VUCA approach would be used in the business world and the complex world of interpersonal relations. One early adopter cited VUCA as appropriate to a world that is "crazy out there!"

Volatility refers to the spread of change within a dynamic setting. Often the speed of this change exceeds expectations and ensnares the naïve or unprepared like a lone swimmer caught in an undertow and dragged down into the ocean's murky depths.

Uncertainty is a semi-confused state resulting from a lack of predictability, incomplete information, or imperfection.

Complexity refers to currently unknown interconnectedness and intricacy resulting from situations with numerous variables.

Ambiguity is the absence of clarity in situations that lend themselves to multiple interpretations.

VUCA shares many of the qualities of the human brain. Neurons are highly connected (a single neuron interacts with ten thousand or more other neurons); normal interactions between neurons lead to behaviors that cannot be predicted or controlled; and no neuron or group of neurons occupy center stage, so to speak. Rather, control is distributed over thousands of subsystems with the final behavior emerging along uncertain, complex, and volatile pathways. VUCA implies working with events in a world that, by their nature, are difficult to predict or control. Changing any of the four VUCA components may lead to novel, unexpected, positive, or even catastrophic consequences.

Both hyperobjects and the VUCA approach increase our ability to simultaneously envision multiple subjects, objects, and events from multiple vantage points. We can—indeed we must—develop the mental skills required to envision multiple intersecting points of view, things not so much seen as inferred, and with an increasing emphasis on thinking polyphonically by considering multiple contributors.

In this way we can discover new and surprising relationships, such as the contribution of war to global warming, as we will discuss in a moment.

But thinking polyphonically won't be easy. Reams of research on attention support the view that the brain operates at its best when it deals with only one thing at a time. Since plasticity is the brain's underlying principle of operation, we can try—and must succeed in—changing our brain to process more than one thing at a time. Otherwise the brain will not be able to manage the extraordinary twenty-first-century challenges that it faces.

As an example, look again at Diagram 1 at the beginning of this book, "Key Influences on the Twenty-First-Century Brain." While each of the contributing influences can be thought about one at a time, it's better to look at the diagram and emphasize the mutual influences of each contribution on the other: heat leading to loss of cognitive sharpness, which leads to susceptibility to misinformation, especially via the internet and social media. Employing the polyphonic thinking style stimulates awareness of previously unrealized contributions. Take the effects of war, for example.

Sunil S. Amrith, history professor at Yale University, speak in his book *Burning Planet* of wars as drivers of planetary harm, "In pursuit of empire and domination, of territorial conquest or racial and religious supremacy, wars stand as a stubborn driver of planetary harm."

Whatever the gains that may result from the current efforts to reduce the world's carbon footprint, it all comes to nothing if the wars around the world continue.

Amrith points to the Russia-Ukraine War and the war in Gaza as examples: "The first two years of the Russian war on Ukraine have generated an estimated 175 million metric tons of carbon dioxide emissions—considerably more than Bangladesh, a country of more than 171 million people, produces annually from fuel combustion."

At the time of the construction of the "Key Influences on the Twenty-First-Century Brain" diagram at the start of the book, I had not thought of a war as a major contributor to environmental destruction extending beyond the limited landmass on which a particular war is fought.

War must be included as a contributor to global harm not only for environmental reasons but because "it warps the moral and political imagination. It dims our empathy for one another and kills our ability to imagine the interdependence with the other species of life that share this earth," wrote Amrith in a *New York Times* op-ed. Incorporating the effects of war on climate change is an example of the polyphonic thinking for which we must strive.

In addition, our understanding of the known contributors to the twentieth-first-century disorder and disarray are turning out to be more dynamic and threatening than we initially realized. During the weeks I was finishing this book, a 2025 study on air quality by the American Lung Association was released; it found that 25 percent more people in the United States are inhaling unhealthy air than only a year earlier in 2024. Another study published in *Science Advances* revealed that microscopic pollution containing fine particles about a thirtieth the width of a human hair can pass indoors. As a result, the official advice to remain indoors during times of air pollution is probably underestimating the percentage of the polluted air that travels inside the home.

Another example: With the expansion of wildfires into historically unusual areas (the Carolinas), the science of air quality determination has become a national—not just a California—challenge. Now damage to the lungs of people living throughout the nation is possible. The requisite knowledge needed to determine the best path to follow has become so extensive that even the collaboration of a cadre of experts is insufficient to enable us to wend our way through the bramble bush of environmental challenges.

Plastics are also turning out to be a greater hazard than experts initially realized. While it's known that plastics contribute about 5 percent of the total greenhouse gas emissions, the effects of micro plastics in the ocean and soil continue to be underappreciated. They disrupt the natural cycles of plankton and other organisms that extract carbon dioxide from the atmosphere and cool the planet.

So on those two issues alone (air quality and plastics), increases in their potential lethality serve as indicators that we must "up our game."

(5) A sensible solution

While writing this book, I experienced frequent surprises at interconnections that would have probably been obvious to someone trained in the appropriate field: Chemistry as the basis for the gasoline-plastics nexus is one example. Indeed, what's needed are specialists in varying fields working together in a collaborative manner like those forensic experts at Quantico homing in on the identity of a serial killer. So far so good. In regard to explaining the interactions of the separate factors discussed in this book, more is needed than just experts.

A widely perceived failure by those considered "experts" has led in recent years to a loss of public confidence. The off again/on again advice about wearing masks during the COVID pandemic, for example, sowed confusion and resentment among the general public. As a result, a significant portion of the population now takes advice from people with no medical training. While the majority of people recognize the wisdom of following the advice of experts with professional training—there hasn't as yet emerged a run on the self-repair of cars, or the self-installation or repair of electric circuits within the home—a significant number of people think of themselves as the best judges concerning medical matters, such as the need for vaccines or other countermeasures aimed at conquering infectious diseases. They are encouraged in this belief by public health figures, who either downplay the need for vaccines and other countermeasures, or put a nix on them all together. So what would be a possible solution to these and the other problems described in this book?

Sure, it would be great to somehow find a single person with the ability to assimilate all of the knowledge from such diverse fields as meteorology, chemistry, biochemistry, thermodynamics, information science, glaciology, volcanology, hydraulic engineering, atmospheric engineering, tropical medicine, sociology, jurisprudence, psychiatry, demography, statistics, diaspora studies, botany—the list could go on. But no individual or group of individuals could possibly master all the knowledge required. So where do we go?

Let's start with a simple premise: Since we are all subject to the environmental challenges of the twenty-first century, our best chance of success depends on harnessing the knowledge of every one of us, even those who at first may seem improbable contributors toward successful solutions.

The idea that only the erudite with advanced academic degrees have anything worthwhile to say while people with more a humble background and education aren't worth listening to is contradicted by Wikipedia, which was established early in this century

(January 10, 2001) by Jimmy Wales and Larry Sanger. The name Wikipedia was suggested by Sanger as a playful combination of *wiki* (a Hawaiian word for "quick") and encyclopedia.

In just under twenty-five years this reference work has demonstrated that solving problems is best done by inviting anyone deeply interested in a particular twenty-first-century challenge to provide their own ideas about how the problem may be solved.

At first it seemed hard to believe that anything worthwhile could result from contributions on a given subject by people lacking specialized knowledge on that subject. In fact, the requirement that "no central organization" would be in charge was off-putting to some people upon first hearing about Wikipedia. It seemed a matter of common sense that only people with detailed knowledge of a subject, buttressed by a string of academic credentials, would be in a position to offer any useful insights.

Actually this assumption was exactly what Jimmy Wales had anticipated. On many occasions he credited his thinking about how to manage the Wikipedia project to an explosively original essay, "The Use of Knowledge in Society," published in 1945 in the *American Economic Review* by the Austrian-British academic economist Friedrich Hayek. Hayek's main point in his short essay of little over a dozen pages was that knowledge never exists in a concentrated or integrated form "but solely as dispersed bits of incomplete and frequently contradictory knowledge, which all of the separate individuals possess."

Among Hayek's other points: "We have to deal with the unavoidable imperfection of man's knowledge and the consequent need for a process by which knowledge is constantly communicated and acquired" and "Practically every individual has some advantage over all others in that he possesses unique information of which beneficial use might be made."

Read again and think for a moment about Hayek's second point. *The truth of that conviction is what can save us.*

In response to the Hayet-inspired insights of Jimmy Wales, Wikipedia took off with the efficiency and beauty of a magical golden bird taking flight into an infinite turquoise sky. In 2001, a hundred articles on September 11 alone appeared in the English version of Wikipedia. By 2003, Wikipedia contained a hundred thousand total articles on widely varying topics. A decade later the print edition of the English Wikipedia consisted of a thousand volumes containing over 1.1 million pages. By 2015, there were a total of thirty-six million articles across what was already 291 language editions. This made possible ten billion global page views with around 495 billion unique visitors every month, including eighty-five million visitors from the United States alone.

For a time, the success of the internet search engines seemed to eclipse, for some people, the performance of Wikipedia. But there are key differences between Wikipedia and the internet.

The Internet	Wikipedia
1. The single biggest source of information on any subject, but lacks coordinated fact checking.	1. The largest and most referenced work in history
2. Commercial for-profit search engines are controlled by profit-seeking entrepreneurs	2. Only slightly less accurate than Encyclopædia Britannica according to an investigation by the journal *Nature*.
3. Widely varying levels of accuracy result in misinformation and disinformation, which, thanks to the absence of fact-checking makes the internet essentially useless.	3. Uses a small army of unpaid fact checkers (known as Wikipedians), currently estimated at over 200,000 to check entries for accuracy.
	4. Advertisement-free and non-profit; does not generate revenue from its content or users.

What's needed now is a kind of specialized Wikipedia aided by AI and devoted to the challenges and threats with which our twenty-first-century brain is currently wrestling. Added to the traditional scientific sources, we will need data drawn from such diverse humanistic sources as philosophy, ethics, logic, history, geography, humanities, and social sciences. But that is not enough. What's missing are practical suggestions based on experiences outside of the laboratory or the classroom.

As with the original Wikipedia, the new proposed specialized Wikipedia should be based on two principles: (1) Information is spread throughout the population with each individual familiar with only a small part of the whole picture. (2) There will be no organization responsible for assigning the submissions. Everyone and anyone will be free to provide contributions in the form of ideas and suggestions. Corrections or emendations of the contributions of any person who was to contribute will be made by the community of coauthors (corresponding to the original Wikipedians).

The twenty-first-century brain—the connectomic brain as we now understand it—is the perfect organ for discovering the relationships and interconnections of forces unique to twenty-first-century problems where:

1. Nothing is localized.
2. Interconnections can be discovered and new information is uncovered quickly and assimilated, leading to the discovery of new interconnections.
3. Solutions to the challenges discussed in this book must be considered as part of a larger dynamic process. Just as we can never confidently declare that a particular cell or network within the brain isn't asserting some influence on other specific brain areas, we have to remain open to

discovering unknown, surprising, unappreciated, and even shocking influences by contributing factors scattered throughout the world.

It's urgent that we get started on this project. All of us need to think *now* about the problems, put our thoughts into words, and share them with others, thereby coming up with solutions either formally—the dedicated Wikipedia I've described—or informally by word of mouth. Challenge yourself to come up with insights and suggestions, and work with them in the Notes section provided at the end of the book. Doing this is of inestimable importance since such a cooperative venture offers us the best hope both for our survival and the health of the twenty-first-century brain.

Acknowledgments

Contributors and contributions are acknowledged under "Sources Consulted."

Thanks to John Wylie and John Barrer for their suggestions. Special gratitude to Timothy Morton for taking the time to explain to me his concept of hyperobjects. Finally thanks to my assistant Franziska Bening for her hard work and dedication to keeping everything organized and on schedule.

Principal Sources Consulted

Aldern, Clayton Page. *The Weight of Nature: How a Changing Climate Changes Our Brains.* Dutton, 2024.

Amrith, Sunil. *Burning Earth: An Environmental History of the Last Fifty Years.* W.W. Norton & Company, September 2024.

Anderson, Janna, and Lee Raine, "The Future of Truth and Misinformation Online." Pew Research Center, October 19, 2017, https://www.pewresearch.org/internet/2017/10/19/the-future-of-truth-and-misinformation-online/.

Associated Press. "Wildfire smoke may be worse for brain health than other air pollution, dementia research finds." July 29, 2024, https://www.pbs.org/newshour/science/wildfire-smoke-may-be-worse-for-brain-health-than-other-air-pollution-dementia-research-finds.

Barakett, Elysee, and Evan Bush. "Monday was the world's hottest day ever recorded—breaking Sunday's short-lived records." NBC News, July 2024, https://www.nbcnews.com/science/environment/hottest-day-ever-record-monday-sunday-rcna163408.

Bellan, Rebecca, and Dominic-Madori Davis. "The uproar over Vogue's AI-generated ad isn't just about fashion." TechCrunch,

August 3, 2025, https://techcrunch.com/2025/08/03/the-uproar-over-vogues-ai-generated-ad-isnt-just-about-fashion/.

Ben Nafa, Hanan, PhD. "Do bilinguals express different emotions in different languages?" Language on the Move, October 18, 2025, https://www.languageonthemove.com/do-bilinguals-express-different-emotions-in-different-languages/.

Bentham, Jeremy. *The Panopticon*. Hudson Street Press, May 2025.

Berry, Ellen, and Graham Dickie. "Intercepting Self-Harm, but Many False Alerts." *New York Times*, December 10, 2024.

Berwyn, Bob. "Bonn Climate Talks Rife with Roadblocks and Dead Ends." Inside Climate News, June 27, 2025, https://insideclimatenews.org/news/27062025/bonn-climate-talks-roadblocks/.

Bhattarai, Abha, and Rachel Siegel. "Crash unlikely to deal major economic blow but is reminder of broader risks." *Washington Post*, July 21, 2024.

Bommer, Michael. "Dying man spends final weeks creating AI version of himself to keep his wife company." *The Independent*, July 5, 2024, https://www.independent.co.uk/news/uk/home-news/artificial-intelligence-version-death-technology-michael-bommer-b2574963.html.

Booth, William. "Britain settles groundbreaking legal case over girl's death from air pollution." *New York Times*, November 1, 2024.

Boswell, John D. "We face daunting global challenges. Here are eight reasons to be hopeful." *The Guardian*, July 29, 2025, https://www.theguardian.com/commentisfree/ng-interactive/2025/jul/29/global-future-challenges-optimism.

Principal Sources Consulted

Bridle, Janes. *New Dark Age: Technology and the End of the Future.* Verso, 2018, updated 2023.

Buruma, Ian. "The Joys and Perils of Victimhood." *New York Review of Books*, April 6, 1999.

Carney, Sean J. Patrick. "The new art to atomic critique." *Bulletin of the Atomic Scientists*, February 17, 2025, https://thebulletin.org/2025/02/the-new-art-of-atomic-critique/.

Casert, Raf. "EU warns deadly flooding and wildfires show climate breakdown is fast becoming the norm." AP News, September 18, 2024, https://apnews.com/article/eu-climate-floods-wildfires-disaster-8338ec7a0030cc8069800b0e95ed61c9.

Castro-Root, Gabe. "Your Rental Car Get Dinged? You Will, Too, Thanks to A.I." *New York Times*, July 9, 2025.

Chang, Max L. Y., and Irene O. Lee. "Functional connectivity changes in the brain of adolescents with internet addiction: A systematic literature review of imaging studies." *PLOS Mental Health*, June 4, 2024, https://journals.plos.org/mentalhealth/article?id=10.1371/journal.pmen.0000022.

Chapman, Tom. "Nobel Prize winner reveals chillingly simple way AI will 'take over the world.'" Unilad Tech, February 4, 2025, https://www.uniladtech.com/news/ai/geoffrey-hinton-predicts-ai-will-take-over-the-world-979951-20250204.

Cherelus, Gina. "Keep Calm and Carry a Portable Fan." *New York Times*, August 8, 2024.

Christensen, Jen. "Nearly half of Americans live in an area with a failing grade for air pollution, and the problem is only getting worse." CNN, April 23, 2025, https://www.cnn.com/2025/04/23/health/air-pollution-report.

Crow, Robert. "More than 60 scientists issue dire warning that the Earth is careening toward catastrophe: 'Things are all moving in the wrong direction.'" Cool Down, July 9, 2025, https://www.thecooldown.com/green-tech/unprecedented-climate-change-warning-global-temperatures/.

Dajose, Lori. "The brain's processing paradox: Study quantifies the speed of human thought." Medical Press, December 17, 2024, https://medicalxpress.com/news/2024-12-brain-paradox-quantifies-human-thought.html.

De Figueiredo, Madeline. "An Alternative to Reality Has the Ring of Truth." *New York Times*, March 2024.

Delgado, Carla, "Extreme Heat on the Brain: How to Think in the Heat." *Discover Magazine*, July 7, 2021, https://www.discovermagazine.com/mind/how-does-extreme-heat-affect-our-brains.

DePillis, Lydia. "Even as Climate Risks Rise, Short-Term Thinking Prevails." *New York Times*, June 21, 2024.

Diaz, Johnny. "Fauci Is Recovering at Home From West Nile Virus Infection After Hospitalization." *New York Times*, August 2024.

Douthat, Ross. "Come with Me If You Want to Survive an Age of Extinction." *New York Times*, April 20, 2025.

Douthat, Ross. "The Post–Cold War Era Has Ended. The Future Is Up for Grabs." *New York Times*, November 17, 2024.

Dzombark, Rebecca. "Almost Half of Americans Are Breathing Unhealthy Air, Report Finds." *New York Times*, April 24, 2025.

Dzombark, Rebecca. "Some Glaciers Will Disappera Regardless of Temperatures, Study Finds." *New York Times International*, May 30, 2025.

Principal Sources Consulted

Edwards, Benj. "Omnipresent AI cameras will ensure good behavior, says Larry Ellison." Ars Technica, September 16, 2024, https://arstechnica.com/information-technology/2024/09/omnipresent-ai-cameras-will-ensure-good-behavior-says-larry-ellison/.

Elbein, Saul. "Most Americans fear global warming. Here's why few discuss it." *The Hill*, April 2025, https://thehill.com/policy/energy-environment/5252198-climate-change-silence-study/.

Elk, Sarah. "3 Takeaways on the Future of AI from the Nvidia Conference." *Forbes*, April 2025, https://www.forbes.com/sites/selk/2025/04/01/3-takeaways-on-the-future-of-ai-from-the-nvidia-conference/.

Elliott, Rebecca F. "Fossil Fuels Regain Ground After Green Energy Falters" and "Fossil Fuels Regain Wall Street's Favor After Renewable Energy Stalls." *New York Times*, November 19, 2024.

Emont, Jon. "Climate Change Is Coming for the Finer Things in Life." *Wall Street Journal*, June 10, 2024.

Farrow, Ronan. *Surveilled*. directed by Matthew O'Neill and Perri Peltz, HBO Documentary Film, 2024.

Firth, Shannon. "Concerns Continue over Private Equity's Reach Into Healthcare—Rising costs have left many hospitals vulnerable to 'the sway of private equity,' one expert says." Med Page Today, January 2025, https://www.medpagetoday.com/hospitalbasedmedicine/generalhospitalpractice/113613.

Frenkel, Sheera, and Natan Odenheimer. "How Isreal Sent A.I. Into Combat: With Technology Came Fatal Consequences." *New York Times*, April 25, 2025.

Fried, Ina. "1 big thing: Chatbots are learning to die." Axios, December 13, 2024, https://www.axios.com/podcasts/1-big-thing.

Friedman, Lisa. "Offhand Remark on Gold Bars, Secretly Taped, Upends a Life." *New York Times*, July 4, 2025.

Fung, Brian. "How one flawed software update could have such widespread effects—and cost "potentially billions of dollars." CNN, July 19, 2024, https://www.cnn.com/2024/07/19/business/recovery-global-crowdstrike-outage.

Furedi, Frank. *The War Against the Past: Why the West Must Fight for Its History*. Polity Press, 2024.

Gaffney, Austyn. "Study Shows Biggest Fires Have Grown More Intense." *New York Times*, June 25, 2024.

Gelles, David. "A.I.'s Insatiable Energy Use Drives Electricity Demands." *New York Times Business*, July 15, 2024.

Gelles, David. "Confronting Our New Reality: Solutions to the problem of climate change have never been more clear. But the scale of the problem keeps getting bigger." *New York Times*, September 25, 2024.

Gelles, David. "Scientist Wants to Block Sunlight to Cool Earth: Growing Interest in Bold Plan Despite Grave Risks." *New York Times*, September 2024.

Ghani, Faras. "Pakistan slams climate 'injustice' as deadly floods hit country again." Al Jazeera, June 28, 2025, https://www.aljazeera.com/news/2025/6/28/pakistan-slams-crisis-of-injustice-as-deadly-flooding-hits.

Goldstein, Brett J., and Brett V. Benson. "The Era of A.I. Propaganda Has Arrived." *New York Times*, August 11, 2025.

Goode, Lauren. "Deepfakes, Scams, and the Age of Paranoia." *Wired*, May 2025, https://www.wired.com/story/paranoia-social-engineering-real-fake/.

Harrington, Mary. "Thinking Is Becoming a Luxury Good." *New York Times*, July 28, 2025, https://www.nytimes.com/2025/07/28/opinion/smartphones-literacy-inequality-democracy.html.

Harwell, Drew. "She live-streams 24/7. Is it life, or a performance of one?" *Washington Post*, May 4, 2025.

Hayek, Friedrich. "The Use of Knowledge in Society." *American Economic Review*, 1945.

Hennigan, W. J. "Nuclear War Avoided, Again. But Next Time?" *New York Times*, May 15, 2025, https://www.nytimes.com/2025/05/15/opinion/india-pakistan-nuclear-war.html.

Hern, Alex. "Google DeepMind takes step closer to cracking top-level maths." *The Guardian*, July 2024, https://www.theguardian.com/technology/article/2024/jul/25/google-deepmind-takes-step-closer-to-cracking-top-level-maths.

Hiar, Corbin, and E&E News. "Big Banks Quietly Prepare for Catastrophic Warming." E&E News with permission from Politico, LLC, March 2025.

Hill, Kashmir. "Professors Face Student Rancor over Use of A.I." *New York Times*, May 18, 2025.

Hill, Kashmir. "Whisper Sweet Nothings, ChatGPT." *New York Times*, January 19, 2025.

Hill, Samantha Rose. "Big Tech Wants to Profit from the Loneliness It Helps Cause." *New York Times*, July 7, 2025.

Hofstadter, Richards. "The Paranoid Style in American Politics." *Harper's Magazine*, November 1964, https://harpers.org/archive/1964/11/the-paranoid-style-in-american-politics/.

Holtermann, Callie. "Having to Prove That A.I. Didn't Do Your Homework." *New York Times*, May 18, 2025.

Hubler, Shawn. "Impact of Wildfires Felt Far Beyond L.A." *New York Times*, April 17, 2025.

Hunter, Tatum. "Am I hot or not? People are asking ChatGPT for the harsh truth." *Washington Post*, June 1, 2025.

Isaac, Mike, and Nicole Sperling. "Meta in Talks to Use Voices of Celebrities for Chatbots." *New York Times*, August 2024.

Jackson, Rob. "New Fixes for Methane Emissions Could Be a Big Climate Help." *Wall Street Journal*, July 25, 2024, https://www.wsj.com/science/environment/the-best-quick-fix-for-climate-change-curbing-methane-b342b192.

Jessop, Simon, and Valerie Volcovici. "Leaders at climate meetings in New York warn of growing mistrust between nations." Reuters, September 22, 2024, https://www.reuters.com/world/leaders-climate-change-meetings-new-york-warm-growing-mistrust-between-nations-2024-09-22/.

Johnson, Mary M., et al. "Immune impacts of fire smoke exposure." *Nature Medicine*, June 26, 2025, https://www.nature.com/articles/s41591-025-03777-6.

Kaplan, Sarah. "Ecosystems losing ability to absorb CO2, scientists fear." *Washington Post*, April 17, 2025.

Kaur, Anumita. "Some Mass. Residents urged to stay in at night over deadly mosquito virus." *Washington Post*, August 2024.

Kelmar, Patricia. "Was your health insurance claim denied by an algorithm? Thousands are: Did a doctor or an AI-bot deny your health insurance claim? What you need to know." Public Internet Network, July 2023, https://pirg.org/edfund/articles/was-your-health-insurance-claim-denied-by-an-algorithm-thousands-are/.

Kitajima, Mulkey, Claire Brown, and Mira Rojanasaku. "Earth Is Being Cooked at a Quickening Pace." *New York Times*, June 17, 2025.

Klein, Ezra. "Now Is the Time of Monsters." *New York Times*, January 19, 2025.

Kolata, Gina. "A.I. Chatbots Defeated Doctors at Diagnosing Illness." *New York Times*, November 19, 2024.

Kurutz, Steven. "You're Surrounded by Scammers: To engage with the digital world to any degree, via computer or smartphone, is to be a potential mark." *New York Times*, April 21, 2024.

Lakhani, Nina. "Carbon footprint of Israel's war on Gaza exceeds that of many entire countries." *The Guardian*, May 2025, https://www.theguardian.com/world/2025/may/30/carbon-footprint-of-israels-war-on-gaza-exceeds-that-of-many-entire-countries.

Lo, Joe. "Developing countries denounce rich nations' disregard for just transition talks." Climate Change News, September 17, 2024, https://www.climatechangenews.com/2024/09/17/developing-countries-denounce-rich-nations-disregard-for-just-transition-talks/.

Loofbourow, Lili, and Sonia Rao. "Location sharing was cool, until it wasn't." *Washington Post*, August 11, 2025.

Love, Shayla. "The Hot New Luxury Good for the Rich: Air." *New Republic*, February 21, 2024, https://newrepublic.com/article/178452/clean-air-rich-luxury-good.

Luccioni, Sasha. "Generative AI and Climate Change Are on a Collision Course: From energy to resource, data centers have grown too greedy." *Wired*, December 2024, https://www.wired.com/story/true-cost-generative-ai-data-centers-energy/.

Lyngaas, Sean. "White House official spoke to CrowdStrike CEO as government assesses impact of global IT outage." CNN, July 19, 2024.

Madani, Kaveh. "Heat Waves and Wildfires Give Urgency to the Fight to Cut Methane." *Forbes*, June 29, 2024, https://www.forbes.com/sites/kavehmadani/2024/07/29/heat-waves-and-wildfires-give-urgency-to-the-fight-to-cut-methane/.

McCarthy, Erica. "Climate Change Will Be More Severe in 2025." New Atlanticist, November 25, 2008, https://www.atlanticcouncil.org/blogs/new-atlanticist/climate-change-will-be-more-severe-in-2025/.

McGill, Timothy. "Scientists raise red flag over dangerous trend in deadly heatwaves: 'The impacts are devastating.'" Cool Down, June 29, 2025, https://www.thecooldown.com/outdoors/extreme-heat-global-climate-change-report/.

McMillan, Robert, and Sarah Krouse. "A Disney Worker Downloaded an AI Tool. It Led to a Hack That Ruined His Life." *Wall Street Journal*, February 26, 2025, https://www.wsj.com/tech/cybersecurity/disney-employee-ai-tool-hacker-cyberattack-3700c931?gaa_at=eafs&gaa_n=ASWzDAg4BFoXWNaHM9fz2L4qQgn6V7lhmP32hZKAw1YVFe88p7YnqfnkTjT&gaa_ts=684af2ab&gaa_sig=CMFFTzXy39EV3wUdGyz3LpXBdqhksK8vVUDmQHyyXALh809iHIt3xhaATE3VJGKDWf7qAoTpZF9MxXBpRN7AJQ%3D%3D.

Mellen, Ruby. "New Study shows that wildfire smoke seeps into homes." *Washington Post*, May 15, 2025.

Meredith, Sam. "State of Freight 'Inadvertent geoengineering': Researchers say low-sulfur shipping rules made climate change worse." CNBC, June 19, 2024, https://www.cnbc.com/2024/06

/19/geoengineering-study-shipping-regulation-made-climate-change-worse.html.

Metz, Cade, and Dylan Freedman. "How Artificial Intelligence Chatbots Like ChatGPT and DeepSeek Reason." *New York Times*, May 5, 2025.

Milman, Oliver. "US banks predict climate goals will fail—but air conditioning firms will thrive." *The Guardian*, April 2, 2025.

Mims, Christopher. "The Summer Is So Hot, Workers Are Wearing High-Tech Ice Packs: New Technologies are making what had seemed like science fiction possible as jobs keep people outside for long periods." *Wall Street Journal*, August 23, 2024.

Mitchell, Amy, Mark Jurkowitz, J. Baxter Oliphant, and Lisa Shearer. "3. Misinformation and competing views of reality abounded throughout 2020." Pew Research Center, February 20, 2021, https://www.pewresearch.org/journalism/2021/02/22/misinformation-and-competing-views-of-reality-abounded-throughout-2020/.

Monroe, Robert. "The Keeling Curve." UC San Diego, Scripps Institution of Oceanography, April 3, 2013.

Morrone, Megan. "Deepfaked after death." Axios, August 9, 2025, https://www.axios.com/2025/08/09/deepfake-death-reincarnation-ai-avatar.

Morton, Timothy. *Hyperobjects: Philosophy and Ecology after the End of the World*. University of Minnesota Press, 2013.

Myers, Steven Lee, and Stuart A. Thompson. "A.I. Is Starting to Wear Down Democracy." *New York Times*, June 26, 2025, https://www.nytimes.com/2025/06/26/technology/ai-elections-democracy.html.

Nelken-Zitser, Joshua. "This will be the largest IT outage in history, bringing Y2K fears to reality, web-security expert says." *Business Insider*, July 19, 2024, https://www.businessinsider.com/global-it-outage-y2k-24-years-later-crowdstrike-cyber-expert-2024-7.

Nicas, Jack. "The Internet's Final Frontier: Remote Amazon Tribes of Brazil." *New York Times*, June 2, 2024.

NIH Research Matters. "Wildfire smoke exposure and dementia risk." National Institutes of Health, December 18, 2024, https://www.nih.gov/news-events/nih-research-matters/wildfire-smoke-exposure-dementia-risk.

Niranjan, Ajit. "'A war of the truth': Europe's heatwaves are failing to spur support for climate action." *The Guardian*, July 4, 2025, https://www.theguardian.com/environment/2025/jul/04/europe-heatwaves-failing-support-climate-action.

Olson, Parmy. "ChatGPT's Mental Health Costs Are Adding Up." Bloomberg, July 4, 2025, https://www.bloomberg.com/opinion/articles/2025-07-04/chatgpt-s-mental-health-costs-are-adding-up.

Orwell, George. *Nineteen Eighty-Four* (also published as *1984*). Secker & Warburg, June 8, 1949.

Ovide, Shira. "How to spot an AI video? You can't." *Washington Post*, August 12, 2025.

Pappas, Georgios. "A more common enemy: How climate change spreads diseases and makes them more dangerous." *Bulletin of the Atomic Scientists*, October 28, 2024, https://thebulletin.org/2024/10/a-more-common-enemy-how-climate-change-spreads-diseases-and-makes-them-more-dangerous/.

Parks, Tim. "Literature in Laputa." *Times Literary Supplement*, February 14, 2025.

Passy, Jacob. "Forget Crowded Airport Lines/Worry About Thunderstorms and Derechos This Summer." *Wall Street Journal*, June 4, 2024, https://www.wsj.com/us-news/climate-environment/summer-travel-storms-el-nino-derecho-hurricanes-6a8959df?gaa_at=eafs&gaa_n=ASWzDAiKF85_K1ym1brp2tDKmhN4Hiu2SQ_XYoxVH6dh_yAINenyyYYq9Qg_&gaa_ts=689a39db&gaa_sig=uOWBkc3opaw45EEz3aR4LcLuxv56C1777QXsuF8G98oZm3m8BaV47y6XTflHVKdrWH5Q43PDWKuvYeoHE5t_g%3D%3D.

Piller, Charles. *Doctored: Fraud, Arrogance, and Tragedy in the Quest to Cure Alzheimer's*. Atria/One Signal Publishers, February 2025.

Phillips, Aleks. "Anthony Fauci recovering from West Nile virus." BBC, August 2024, https://www.bbc.com/news/articles/c628okxvxpro.

Politi, Daniel, and Natalie Alcoba. "A.I.-Fueled Smear Attack Highlights Argentine President's War on Press." *New York Times*, July 3, 2025.

Pomeroy, Ross. "Can humans purge the bots without sacrificing our privacy?" Freethink, November 30, 2024, https://www.freethink.com/the-digital-frontier/personhood-credentials.

Pribram, Karl H. *The Form Within: My Point of View*. Prospecta Press, February 19, 2023.

Price, Kiley. "Climate Change Is Fueling the Loss of Indigenous Languages That Could Be Crucial to Combating It." Inside Climate News, June 2, 2024, https://insideclimatenews.org/news/02062024.

Rapp, Lauren. "NY governor calls outage 'Unprecedented situation.'" CNN, July 19, 2024.

Restak, MD, Richard M. "Warning: Graphic Images: Footage from a war and the effects on your brain." *The American Scholar* published by Phi Beta Kappa, May 3, 2024, https://theamerican scholar.org/warning-graphic-images/.

Restak, MD, Richard M. *The Brain: The Last Frontier*. Doubleday, July 1, 1979.

Reynolds, Matt. "Science Is Full of Errors. Bounty Hunters Are Here to Find Them." *Wired*, June 21, 2024, https://www.wired.com /story/bounty-hunters-are-here-to-save-academia-bug-bounty/.

Rivera, Nicolas. "Plastic's role in warming may be undercounted." *Washington Post*, May 15, 2025.

Robock, Alan, Owen B. Toon, et al. "How an India-Pakistan nuclear war could start—and have global consequences." *Bulletin of the Atomic Scientists* 75 (no. 6): October 28, 2019.

Rogers, Kaleigh. "What Constant Surveillance Does to Your Brain." Vice, November 2018, https://www.vice.com/en/article /what-constant-surveillance-does-to-your-brain/.

Rogers, Kristen. "How internet addiction may affect your teen's brain, according to a new study." CNN Health, June 5, 2024, https://www.cnn.com/2024/06/04/health/internet-addiction-teen -brain-activity-wellness.

Santos, Juan Manuel. "Countries are neglecting the existential threat of pandemics. Bold leadership is necessary." *Bulletin of the Atomic Scientists*, January 2025, https://thebulletin.org/2025/01 /countries-are-neglecting-the-existential-threat-of-pandemics -bold-leadership-is-necessary/.

Satariano, Adam, Paul Mozur, Kate Conger, and Sheera Frenkel. "Software Update Goes Awry, Causing Chaos Throughout the World." *New York Times*, July 20, 2024.

Schechner, Sam, Gareth Vipers, and Alyssa Lukpat. "Major Tech Outage Grounds Flights, Hits Banks and Businesses Worldwide." *Wall Street Journal*, July 19, 2024.

Schlanger, Zoë. "America's Coming Smoke Epidemic." *The Atlantic*, June 27, 2025, https://www.theatlantic.com/science/archive/2025/06/wildfire-smoke-epidemic/683343/.

Schneider, Gregory S. "Va. Rep. Wexton finds her voice through AI: Tool re-creates sound of her speech as disease limits her ability to talk." *Washington Post*, July 13, 2024.

Sengupta, Somini. "A Quest to Clean the Climate by Using Dirt to Capture Carbon." *New York Times*, August 14, 2024.

Shaw, Gine. "Microplastics Accumulate at High Levels in the Brain." *Neurology Today* 25 (no. 9): May 1, 2025.

Singer, Natasha. "Google Plans to Roll Out A.I. Chatbot to Children." *New York Times*, May 25, 2025.

Smil, Vaclav. *How the World Really Works: The Science Behind How We Got Here and Where We're Going*. Viking, 2022.

Smith, Dana G. "In an Aggressive Heat Wave, Brains React With Similar Hostility." *New York Times*, June 19, 2024, https://www.nytimes.com/2024/06/19/well/mind/heat-affect-brain-emotions.html.

Solman, Paul, et al. "AI and the energy required to power it fuel new climate concerns." *PBS NewsHour*, July 4, 2024, https://www.pbs.org/newshour/show/ai-and-the-energy-required-to-power-it-fuel-new-climate-concerns.

Spratt, David. "Is scientific reticence hindering climate understanding?" *Bulletin of the Atomic Scientists*, March 12, 2025.

Spring, Jake. "State Department fires remaining employees who worked on climate change." *Washington Post*, July 11, 2025.

Stack, Megan K. "China Got Used to Surveillance. Will We?" *New York Times*, June 28, 2025.

Strobel, Warren P., and Ellen Nakashima. "CIA ramps up its hunt for human intelligence." *Washington Post*, May 29, 2025.

Subbaraman, Nidhi. "Flood of Fake Science Forces Multiple Journal Closures." *Wall Street Journal*, May 14, 2024, https://www.wsj.com/science/academic-studies-research-paper-mills-journals-publishing.

Tabuchi, Hiroko. "Plastic Waste from U.S. Is Rekected by Malaysia." *New York Times*, July 1, 2025.

Thebault, Reis. "Death Valley keeps getting hotter. How do residents survive?" *Washington Post*, August 11, 2024.

Thomes, Tobi. "Internet addiction alters brain chemistry in young people, study finds." *The Guardian*, June 19, 2024, https://www.theguardian.com/technology/article/2024/jun/04/internet-addiction-alters-brain-chemistry-in-young-people-study-finds.

Tiku, Nitasha. "That chatbot companion? It may lead you astray: 3Researchers just beginning to grapple with pros, cons of chatbot relationships." *Washington Post*, June 2025.

Toon, Owen B., et al. "Rapid Expansion of Nuclear Arsenals by Pakistan and India Portends Regional and Global Catastrophe." *Science Advances* (no. 5): 2025.

Treisman, Rachel. "The Doomsday Clock has never been closer to metaphorical midnight. What does it mean?" NPR, January 2025, https://www.npr.org/2025/01/29/nx-s1-5279204/doomsday-clock-2025-history.

Tufekci, Zeynep. "The Ocasio-Cortez Deepfake Was Terrible. So Was the Proposed Solution." *New York Times*, August 13, 2025.

Turco, Richard P., and Owen Brian Toon. "The fires of Hiroshima and Los Angeles: Apocalypse redux." *Bulletin of the Atomic Scientists*, February 12, 2025, https://thebulletin.org/2025/02/hiroshima-and-los-angeles-compared-apocalypse-redux/.

Twilley, Nicola. *FROSTBITE: How Refrigeration Changed Our Food, Our Planet, and Ourselves*. Penguin, June 2025.

Vigdor, Neil, and Aishvarya Kavi. "Landslide from Fractured Glacier Buries Evacuated Village of 300 in Swiss Alps." *New York Times*, May 28, 2025.

Waldman, Annie, and Sharon Lerner. "NIH Ends Future Funding to Study the Health Effects of Climate Change." ProPublica, March 26, 2025, https://www.propublica.org/article/nih-funding-climate-change-public-health.

Wallace-Wells, David. "The Flooding in Texas Is the Future We Need to Prepare For." *New York Times*, July 13, 2025.

Wallace-Wells, David. "The World Is Now Unavoidably Toxic." *New York Times*, August 10, 2025.

Wang, Vivian. "What Should You Trust? Not Much, Chinese Intelligence Says." *New York Times*, September 4, 2024.

Ward, Ashley. "Extreme Heat Is Breaking America." *New York Times*, June 25, 2025.

Weise, Karen. "Amazon Unveils Alexa+, Powered by Generative A.I." *New York Times*, February 2025.

Wilkins, Joe. "This May Be the Most Terrifying Sign of Global Warming Yet: Follow the money." Futurism, April 6, 2025, https://futurism.com/global-warming-banks.

Williams, Ashley R., and Michelle Watson. "High mosquito-borne encephalitis risk prompts Massachusetts town to close parks, fields at night." CNN. August 24, 2024, https://www.cnn.com/2024/08/24/health/plymouth-ma-encephalitis-risk-closures.

Woollacott, Emma. "Climate Change Deniers Are Switching Tactics." *Forbes*, June 24, 2025, https://www.forbes.com/sites/emmawoollacott/2025/06/24/climate-change-deniers-are-switching-tactics/.

Yang, Maya. "Series of US mass shootings brings weekend of death and mayhem." *The Guardian*, June 23, 2024, https://www.theguardian.com/us-news/article/2024/2024/jun/23/mass . . . ings-new-york-alabama-missouri-ohio?CMP=oth_b-aplnews_d1.

Zhong, Raymond. "Study Finds Alaskan Ice Field Melting at an 'Incredibly Worrying' Pace." *New York Times*, July 2, 2024, https://www.nytimes.com/2024/07/02/climate/alaska-juneau-icefield-melting.html.

Notes

Notes

Notes

Notes

Notes

Notes